Waste: A Reader About the Global Routes of Rubbish

H. R. Decker, Nur dumme Fische sterben in verschmutzten Flüssen, 1974, Artist's book, offset and silkscreen on paper; Collection Museum Ostwall at the Dortmunder U, ©VG Bild-Kunst

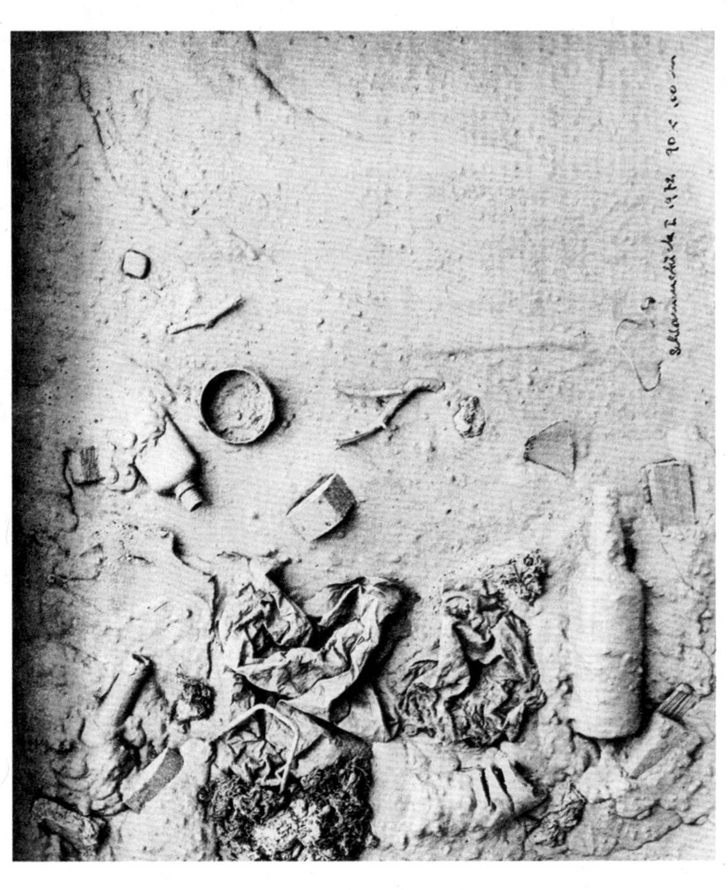

FOREWORD

Regina Selter
Director, Museum Ostwall at the Dortmunder U

4 / 5

HA Schult, Environment as part of the exhibition HA Schult: Der Macher at the Museum Ostwall at the Dortmunder U, Dortmund 1972

In the summer of 1978, at what was then Museum am Ostwall, artist HA Schult put an extraordinary environment on display: Piles of scrap wood, old pieces of furniture, and electrical appliances were distributed around the central atrium, arranged in small clusters or staged as individual elements. The large pieces of trash on show came from people's homes in Dortmund.

An article on the exhibition that appeared in the *Tagesspiegel* newspaper described the large-scale installation as a "non-space," saying that it was ultimately an assortment of objects that were not typically destined for the rooms of a museum and were broken and dirty to boot. Then, as now, Museum Ostwall's programming and sense of identity centered on the connection between art and life and an engagement with everyday experiences. HA Schult's temporary installation reflected an aspect of day-to-day life that is mostly invisible and disappears into dumpsters and garbage trucks, refuse tips, and incinerators.

In the time since HA Schult presented his Environment at Museum am Ostwall, the pathways along which waste travels have become increasingly complex: Today, household refuse, e-waste, textiles, and industrial waste move everywhere around the globe. For most consumers, it is almost impossible to reconstruct the course this waste takes, or to understand where it winds up in the end. Yet its economic importance can hardly be underestimated, as it is sold to other countries both legally and illegally, giving rise to a trading network and a web of relationships of dependency that are invariably shaped by the balance of global power. This book seeks to delve into the complexity of these relationships and give a broader sense of the background to the pathways that waste follows around the world. The various contributions act as a bracket that connects art concepts with societal developments.

Artists continue to focus on the theme of waste, not only because of its material quality but also as a means to examine and expose global connections and dependencies. Just as important, they show how utopias can emerge from waste and how distinctive new worlds may be created from other people's leavings. The artists represented in *Waste: An Exhibition About the Global Routes of Rubbish* follow the pathways traveled by waste, leveling accusations, conducting analyses, and ultimately asking what kind of renewal can begin when the world is flooded with waste.

I would like to express my thanks to all the artists and lenders involved in the exhibition, to the authors of this book, and to the publisher Spector Books for its singular design concept and for the inspiring process of collaboration. I would like to express my special thanks to Hannes Drißner for his unique design of this reader. I also thank Christina Danick and Michael Griff, the curators of the *Waste* exhibition and editors of this book. The perspectives they reveal in the exhibition and publication give a concrete sense of the interrelationship between art, life, aesthetics, ecology, and society.

NOKIA

WAYS OF WASTE:

ON THE GLOBAL ROUTES OF RUBBISH

Christina Danick &
Michael Griff

A. J. Weberman had already been fixated on Bob Dylan and his work for several years when he began rummaging through the musician's domestic trash. Weberman reproached Dylan for no longer writing politically motivated music and for turning instead to the mainstream. Using a method of "Dylanology" he had come up with himself, the author and activist analyzed the texts of Dylan's songs, which subsequently led to the publication of the 536-page dictionary *Dylan to English*.[1] Dylan's trash served as material for his research, whose purpose was to allow inferences to be drawn about the singer's private life and, far more important for Weberman, to offer clues to reading his lyrics. Weberman called this "garbology"—a term that is commonly used today for the study of garbage in relation to research into modern societies.

It is debatable whether this branch of research actually owes its name to Weberman. However, there is one insight that can be garnered from his research: We are what we discard. What we consume and dispose of is indicative of how we live. It is for good reason that numerous archaeological finds that today shed light on earlier ways of life were once rubbish.[2] It is not only the rubbish as such that is revealing but rather the way in which we deal with it: In many cases, this involves blocking it out in our everyday lives. In Europe and North America, trash is discarded, picked up, and made to vanish out of sight and out of mind. It stands out all the more, then, when the refuse collectors go on strike, and garbage piles up on the streets. This is set against the enormous amount of trash that is produced: Around the world, two billion tons of municipal waste are disposed of each year, and by 2050 this amount is expected to have increased by 56 percent.[3] Without even bringing industrial waste into the equation. Some 460 million tons of plastic waste are produced worldwide, and here too the numbers are following an upward trend. The most recent negotiations in August 2025 to secure a global agreement on plastics have proved fruitless.[4]

There are numerous depictions of garbage to be found in modern and contemporary art. This also includes taxonomies of trash. Since the late 1950s, in particular, there has been a visible trend, as art historian Amanda Boetzkes argues, for artists to make trash visible and put it on display in exhibitions—on the one hand depicting it in paintings and graphic works, and later in films and photographs as well, while, on the other, using it as artistic material. Representations of refuse and refuse collectors can already be found in the art of the late nineteenth century, such as Edouard Manet's *Le Chiffonier* (1869), and of the first half of the twentieth, such as Julian Trevelyan's *Rubbish May Be Shot Here* (1937). Both artists addressed, at this early stage, the social issues associated with waste disposal. In contemporary art, however, a trash aesthetic has developed that highlights a fraught dynamic: While the aesthetics and materiality of garbage prompt us to reflect—on waste and its seeming invisibility and on the global pollution it causes—they are themselves a

A Edouard Manet, Le Chiffonier, 1869

symbol of these problems, which are engendered by our overconsumption.

The journey that waste takes begins with its disposal. It is sorted, sold, shipped, illegally disposed of, dumped, stored, piled up, and in some cases put to a new use. The paths it follows lead around the globe but remain invisible to those who have started it on its way with a trip to the trash can. Yet it ends up somewhere, and people have to live with and manage the waste there—and anywhere where their own waste is not sold, incinerated, or recycled. They can scarcely avoid it. We know that waste constitutes a burden not just for humans but for all living creatures. We have all seen images of plastic waste in the oceans.[5] Only about a fifth of the waste accumulated around the globe is recycled, and for plastics the numbers are even lower. However, the remainder is not necessarily incinerated or dumped under the stipulated conditions, which makes it difficult to track the actual amount of waste that is treated worldwide in one of these three ways. In many countries, regulations

B Arman, Untitled, 1972

C Julian Trevelyan, Rubbish May Be Shot Here, 1937

are less strict than they are in the industrialized nations of the West. Here, too, the global economic gradient plays a role: There are huge costs involved in disposing of waste properly or getting it out of the country. Even the trade in waste does not always follow legal channels: Somewhere between 15 and 30 percent of the waste exported from the EU is moved illegally, often shipped to Southeast Asia.[6] Whether legal or illegal, the routes waste takes are not always direct; it is often sold to brokers, which makes it even more difficult to track.

The emergence of international environmental movements, the founding of Greenpeace and Friends of the Earth International (both in 1971), the European Conservation Year (1970), and the publication of *The Limits to Growth* report for the Club of Rome (1972) focused attention on global waste production and its ecological consequences. Refuse also became a topic in cultural and social sciences. Mary Douglas, for example, was one of the first social anthropologists to apply herself to the phenomenon of waste in her definitive work *Purity and Danger* (1966). According to Douglas, waste is "matter out of place," that which does not fit into the particular scheme of things: Since this system differs according to the geographical, cultural, and religious context, what we characterize as waste also differs.[7] Michael Thompson's *Rubbish Theory* (1979) saw waste as embedded in a value-added cycle defined in terms of the relationship between waste and durable and transient objects, which is however subject to constant shifts. Douglas and Thompson noted that waste eludes any clear definition; it is, instead, context dependent.

This becomes even clearer when we look at the global level, which Douglas and Thompson disregarded by and large: Waste changes in value as it travels along its path. A washing machine that has been disposed of in Germany, that has become worthless there, in other words, is shipped to Nigeria, where it—or rather its individual components—are reused and, most importantly, sold. The same applies to a cell phone that is thrown away in China: The rare earths in it are recovered in Ghana, a process known as urban mining. This revaluation and recycling have little potential to emancipate those involved in the process; it is induced, first and foremost, by an economic imperative and carried out in conditions that put a burden on society and on people's health. Global economic power relations, which are frequently based on colonial continuities, are reflected in the way waste is handled.

Since the late 1950s, trash art has been in vogue. Marcel Duchamp, for example, started using found objects in his works to allow him to break free of the classical canon of materials: Trash was interesting, in aesthetic terms above all. In the wake of this, the work of the New Realists constituted a transition to a new relationship with trash in art. In the late 1950s, artists such as César and Arman were already using trash as a material to draw attention, as a critique of consumerism, to the throwaway culture of Western societies. Arman, whose *Poubelles* were the first works to put accumulations of trash in glass cases and present them as works of art, saw himself as a "witness to this society," one who had "always been intensely concerned with the pseudo-biological cycle of production, consumption, and destruction."[8] *Waste: An Exhibition About the Global Routes of Rubbish*—the show at the Museum Ostwall at the Dortmunder U that has occasioned the publication of this book—sheds light on this transition to a sociopolitical engagement with consumption.

As public awareness of the environment and waste continued to grow, artists voiced increasingly explicit criticism in their work about how these matters were being handled. Action artists such as HA Schult and Allan Kaprow

D HA Schult, Situation Schackstraße, 1969/70

Allan Kaprow, Transfer (for Christo), 1968

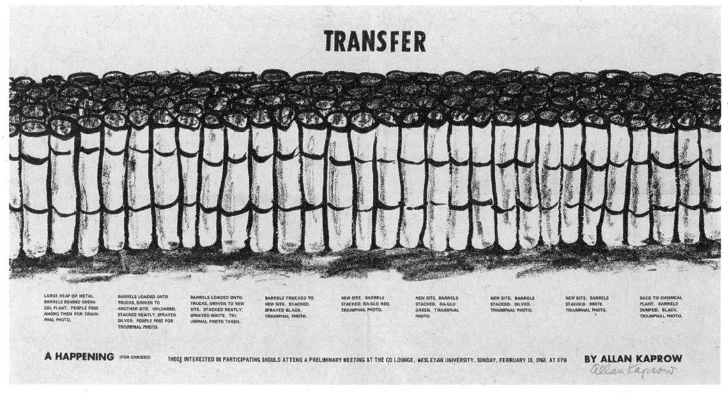

filled different locations with trash or had oil drums taken from place to place to draw attention to the overconsumption of Western societies. From the 1970s on, the disposal and storage of waste—mostly in the form of landfills—played an increasingly important role in the work of artists such as Nancy Holt and Agnes Denes. They demonstrated how these places could be used in new and ecologically sensible ways, thereby breaking, as Amanda Boetzkes argues, with the stigma attached to waste dumps.[9] Of key importance here was the work of Mierle Laderman Ukeles, whose *MANIFESTO FOR MAINTENANCE ART, 1969! Proposal for an exhibition: "CARE,"* called, back in 1969, for an artistic practice focused on care work and the issues of maintenance and sanitation work. The works of Denes, Holt, and Laderman

MANIFESTO!

MAINTENANCE ART -- Proposal for an Exhibition

"CARE"

Mierle Laderman Ukeles

I. IDEAS:

A. The Death Instinct and the Life Instinct:

The Death Instinct: separation, individuality, Avant-Garde par excellence; to follow one's own path to death--do your own thing, dynamic change.

The Life Instinct: unification, the eternal return, the perpetuation and MAINTENANCE of the species, survival systems and operations, equilibrium.

B. Two basic systems: Development and Maintenance. The sourball of every revolution: after the revolution, who's going to pick up the garbage on Monday morning?

Development: pure individual creation; the new; change; progress, advance, excitement, flight or fleeing.

Maintenance: keep the dust off the pure individual creation; preserve the new; sustain the change; protect progress; defend and prolong the advance; renew the excitement; repeat the flight.

show your work--show it again
keep the contemporaryartmuseum groovy
keep the home fires burning

Development systems are partial feedback systems with major room for change.

Maintenance systems are direct feedback systems with little room for alteration.

Ukeles and their approach to waste have undoubtedly influenced contemporary artists like Tejal Shah.

By the 1990s, if not before, as the third wave of globalization got underway, there was an increasing awareness of global economic nexuses and power dynamics in artistic discourse. This was also related to the ramifications of environmental pollution and waste production on a global scale. Nicolás García Uriburu was one of many artists who were now involved in the Greenpeace protest campaigns. Known worldwide for his *Coloraciones*, in which he dyed bodies of water fluorescent green to draw attention to how polluted they were, he took part in numerous other Greenpeace actions and designed banners for the environmentalist organization.

Today, numerous artists are looking at the complex global connections underlying waste production and the waste trade. In works ranging in style from documentary and investigative to speculative and utopian, they focus on regional stories and specific cases that in turn have a great deal to say about global linkages. Using intensive research as the basis for her projects, Ana Alenso examines resource cycles, bringing the circulation of raw materials into her work: Water, air, and even sludge are constantly being pumped through her installations. For the exhibition, Alenso has expanded her ongoing research on gold mining, the use of the metal in cell phones and smart devices, the urban mining of gold, and the global gold

F Mierle Laderman Ukeles, MANIFESTO FOR MAINTENANCE ART, 1969! Proposal for an Exhibition "CARE", 1969 (detail: page 1 of 4)

trade, translating it into an associative, speculative installation: *Obsolete Swing*. The photographs from the installation can be found throughout this reader as a series of images.

In the ecological, economic, and sociopolitical debate on waste, Michael Thompson's assessment still seems to be valid—"Everything can be waste, and everything can become waste," as Roman Köster encapsulates it. At the same time, Köster says, this approach provides few concrete pointers when the consequences of waste are to be measured. According to Köster, this "broad" concept of waste runs the risk of losing sight of its material dimensions.[10] Exhibitions on trash in modern and contemporary art in a sense reflect this dilemma: Focusing on the use of rubbish as an artistic material brings to the fore the question of materiality in aesthetic terms, rather than in terms of the global ecological consequences caused by the production of waste material, by over-consumption, and by the global routes along which rubbish is traded.[11] The exhibition *Waste* thus applies a "narrow" concept of it and looks squarely at its material dimensions and the global economic consequences resulting from it. As the exhibition was being developed, this involved collaborating with our "critical friends" Eva Rost, Peter Emorinken-Donatus, and Roman Köster—experts who reflected on and influenced the concept and realization of the exhibition by introducing different perspectives, viewing it through the lens of cultural history and from an activist and pedagogic standpoint.[12] The involvement of the Dortmund Climate Alliance provided a platform for local initiatives and groups campaigning for ecological issues in the city, while also generating another layer of information in the exhibition.

Waste: A Reader About the Global Routes of Rubbish also follows the logic of a "narrow" concept of waste. It brings together essays from various disciplines that amplify and delve deeper into the themes of *Waste*, often incorporating the works on display in the exhibition. They also present an overview of current debates on the global routes taken by trash. The book does not merely serve as a complement to the exhibition; it is also a full-fledged publication in its own right.

As one of the project's "critical friends," author and historian Roman Köster ponders the complex question of how waste should be defined, based on his book *Müll: Eine schmutzige Geschichte der Menschheit* (Rubbish: A Dirty History of Humanity), which was an important point of reference in the process of devising the exhibition. Media and cultural scientist Jennifer Gabrys focuses on the material aspects of digital infrastructures and the extent to which these also play a role in seemingly immaterial forms of artificial intelligence. Amanda Boetzkes relates the toxicity of waste to queer bodies and asks how this can be used to imagine potential ways of dealing with waste in the future. Artist Susi Gutsche looks at how waste can be read on the basis of Erwin Panofsky's interpretive model of iconography and iconology. Nedine Moonsamy examines the importance of waste, disposal, and renewal in Afro-futurism. Media and performance scholar Evelyn Wan draws on the example of a Hong Kong dump for electronic waste to show how global and postcolonial dependencies underpin environmental pollution and conditions of poverty in the immediate vicinity of these dumps. Author and philosopher Oliver Schlaudt examines the work in perpetuity of Bitterfeld-Wolfen, once the center of the East German chemical industry, which for a long time also served as a landfill site for waste imported from West Germany. Annabel Keenan expands the scope of this book with an essay on art and activism, specifically looking at Greenpeace's collaborations with artists.

We would like to thank all the artists on whose work the exhibition and this book have been built. They have served as our inspiration and contextual starting point, determining the way we view waste. Our thanks, too, to all the authors of this book for the essays they have contributed, which greatly expand the cosmos in which the phenomenon of waste exists.

Christina Danick & Michael Griff

Christina Danick is head of the exhibition team and curator at the Museum Ostwall at the Dortmunder U. Prior to this, she worked as a curatorial assistant and program dramaturg at Urbane Künste Ruhr. She studied cultural studies, art history, and media studies in Hildesheim, Berlin, and Paris.

Michael Griff works as a curator specializing in community and public engagement at the Museum Ostwall at the Dortmund U. This includes oversight of the MO_Beirat project, the museum's citizens' advisory board. Prior to this, he completed a scientific traineeship at the Dresden State Art Collections. He studied history in Berlin, Freiburg, Dublin, and Bologna.

Notes

1 A. J. Weberman, *Dylan to English Dictionary* (Booksurge, 2005).
2 See Roman Köster, *Müll. Eine schmutzige Geschichte der Menschheit*, (C. H. Beck, 2023), 21.
3 Zoë Lenkiewicz et al., *Beyond an Age of Waste: Turning Rubbish into a Resource*, Global Waste Management Outlook 2024 (UNEP / ISWA, 2024).
4 "Talks on Global Plastic Pollution Treaty Adjourn Without Consensus," United Nations Environment Programme, August 15, 2025, https://www.unep.org/news-and-stories/press-release/talks-global-plastic-pollution-treaty-adjourn-without-consensus.
5 OECD, *Global Plastics Outlook: Policy Scenarios to 2060* (OECD Publishing, 2022).
6 "5 Things You Should Know About Waste Trafficking in Southeast Asia," United Nations Office on Drugs and Crime, last accessed November 26, 2025, https://www.unodc.org/unodc/environment-climate/webstories/waste-trafficking-southeast-asia.html
7 See Mary Douglas, *Purity and Danger* (Routledge, 2002), 36.
8 Henry Martin, *Arman: Or Four and Twenty Blackbirds Baked in a Pie; Or Why Settle for Less When You Can Settle for More*, (Abrams, 1973).
9 See Amanda Boetzkes, *Plastic Capitalism: Contemporary Art and the Drive to Waste* (MIT Press, 2019), 79–84.
10 See Roman Köster, *Müll. Eine schmutzige Geschichte der Menschheit*: (C. H. Beck, 2023), 14.
11 For a positive example, see Sandra Beate Reimann, *Territories of Waste: Über die Wiederkehr des Verdrängten* (Museum Tinguely, 2022).
12 Part of this publication is a conversation with the "Critical Friends" of the exhibition, see 153–59.

Illustrations

A Edouard Manet, Le Chiffonnier, 1869, oil on canvas, Norton Simon Museum, Pasadena
B Arman, Untitled, 1972, Rubbish cast in polyester, 100 × 50 × 10 cm, The Arman Marital Trust, Corice Arman
C Julian Trevelyan, Rubbish May Be Shot Here, 1937, graphite, ink, watercolor, collage on paper, Tate London, © Bridgeman Images, Reproduction: Tate London.
D HA Schult, Situation Schackstraße, 1969/70, silkscreen print on paper, Collection Museum Ostwall at the Dortmunder U, © VG Bild-Kunst, Bonn 2025
E Allan Kaprow, Transfer (for Christo), 1968, offset on paper, © Allan Kaprow Estate; photo: Jürgen Spiler
F Mierle Laderman Ukeles, MANIFESTO FOR MAINTENANCE ART, 1969! Proposal for an exhibition: "CARE," 1969, written in Philadelphia, PA, October 1969, Four typewritten pages, each 8 ½ × 11 in., detail: page 1 of 4, © Mierle Laderman Ukeles, Courtesy the artist and Ronald Feldman Gallery, New York

8F
CWT.03A
1/4
UE4S_03
9854259

WHAT IS WASTE?

Roman Köster

Waste is one of those aspects of everyday life that is hard to explain. Intuitively, we know what it is—after all, we throw things away without giving it a great deal of thought. If asked why we do this, we would probably explain that they are dirty or no longer of any use. But what does that actually mean? What do we base these assessments on?

It is actually well-nigh impossible to locate any objective properties that turn something into refuse. Admittedly, waste law—as codified for the first time in Germany's 1972 Waste Disposal Act—does contain an "objective" definition of waste: Refuse comprises objects and substances, such as hazardous chemicals, "whose orderly disposal is imperative in order to protect the public good."[1] However, the clarity this apparently creates is only superficial. After all, many hazardous substances can be recycled, and even real dangers need not prevent us from finding a use for things: We can make a necklace from plutonium, for example—we just can't wear it for very long.[2]

Most importantly, though, these kinds of "objective" properties bear absolutely no relation to most of the waste we produce. Cheese packaging doesn't pose any real threat to us, nor does a banana peel—which is why the objective legal definition of waste has been supplemented by a subjective one: Waste is, in addition, whatever people *want* to dispose of.[3] Yet, what this wanting depends on is not specified in any more detail and is certainly not a matter that would lend itself to legal argument.

MAKING SENSE OF DUMPING

The first step in bringing this issue into sharper relief is to examine the act of turning something into trash, of throwing it away. This seemingly harmless act is made even more harmless by repeated talk of things being thrown away "carelessly."[4] This soon spawns complications, as thoughtlessness gives rise to multiple environmental problems. Yet the impression remains that throwing things away is basically an unconscious, thoughtless act.

Why do we discard things in the first place? Setting aside specific value systems, throwing things away serves an organizing function: Cultural sociologist Aida Bosch has shown, for example, that we build up a relationship with the things in our everyday lives that essentially helps us to create a degree of predictability in what we expect.[5] But this can only work so long as we don't cram our immediate environment full of stuff. Befriending things, as Bruno Latour has said, only functions if we keep our friendships to a reasonable number. Otherwise, it will inevitably result in cognitive overload.[6]

This is a problem primarily because, in modern consumer societies, a vast number of things crowd into our everyday lives, including, for, example, everything we buy at the supermarket.[7] This sets limits to repurposing: The first plastic bottle can be made into a flower vase; but there's already a new one the next day, and maybe even two the day after that. This means that we *need* to throw things away if we are not to get lost in the welter of our everyday lives and relinquish the stability of our immediate environment. It is telling that the well-known compulsive hoarding syndrome, whereby people struggle to get rid of things, is not regarded, first and foremost, as a hygiene problem but rather one of social disorganization.[8]

However, this does not provide an answer to the question of *how* we throw things away, of why we part with certain things and hang on to others. A tidying artist like Marie Kondo has shown that, aside from a very few things, our everyday life consists, in its entirety, of potential garbage. Her success turns on the discovery of how easy radical decluttering can make our lives. Just as significant, however, is her subsequent discovery that throwing away too much leads, at some point, to sterility and sensory deprivation.[9]

WASTE AND ORDER

Throwing things away is thus a cognitive feat of organization, in which we class certain things as worthless, while at the same time declaring others to be useful, clean, and worth holding on to. We remove some things from sight and charge others with value. The absence of any functional purpose certainly plays a role here: What are you going to do with a used cheese wrapper? Questions of abundance and scarcity are obviously important, because the more we can avail ourselves of stuff cheaply, the easier it is for us to part with old things. And finally, we should not forget our ideas of hygiene—i.e., our perceptions of what is dirty and clean.[10]

Historically speaking, there have been major changes in all of these aspects over time: Western societies, in particular, have become increasingly affluent. Mass production practices have shifted, with packaging playing a key role in this and now making up a significant amount of household waste. Our ideas about hygiene are completely different from those that prevailed in the Middle Ages. But is that a problem? Value systems change from one historical period to another, without it necessarily being possible to speak of "better" or "worse." And griping about the ramifications of our prosperity can easily smack of arrogance, especially with respect to parts of the planet that have to contend with severe poverty.

However, in sociological debates, for example, the problems associated with the normative aspect of our ideas about dirt and cleanliness are brought into focus. The benchmark here is the 1966 book *Purity and Danger* by British anthropologist Mary Douglas, which is, by some distance, the most frequently cited work in waste research.[11] This is a rather improbable classic of environmental discourse: Douglas focuses, after all, on Old Testament texts for the most part. At the same time, her reflections can be reduced to a comparatively simple common denominator: Cleanliness is a political concept. It can be used to demarcate a realm of clarity, virtue, and order, while everything that lies outside it can be discarded as dirty, dangerous, and abnormal.

If we follow Douglas's reasoning, then we do not produce waste, we *reproduce* it: We have internalized concepts of dirt, cleanliness, and loss of functionality and keep perpetuating them. We become part of hegemonic structures ourselves—when we shudder, for example, at other people not disposing of their waste properly. Hygiene concepts still remain in place for us even once we have realized that they make no sense (e.g., changing our underwear every day).[12]

At the same time, hegemonic structures of this kind are by no means limited just to waste. Rather, the strategies of inclusion and exclusion that turn an object into waste can be observed everywhere in society. This gives waste a completely different epistemic standing. Our approach to it is indicative of how normative ideas come to hold sway and the concrete practices that are tied up with this.

WASTAGE, OVERLAY, DEVALORIZATION

Douglas's reflections are important if we are to understand the social relevance of exploring waste and the reasons it plays such a major thematic role in art. In fact, it is much more than just an unusual material; it has been used *demonstratively* by artists ranging from Kurt Schwitters and Joseph Beuys to El Anatsui.[13] Here, we can identify three important issues relating to communal life in society that can be addressed and critically interrogated through an engagement with waste.

The first thing to mention is the problem of *wastage*: Trash is symbolic of the pathological excesses of an economic system that not only contaminates the air and water but also produces massive surpluses that are subsequently thrown away. Waste streams reveal how capitalism is blatantly failing to adjust production to our needs. Some people speak of a waste economy, and latterly—influenced by the discourse of the Anthropocene, which is to say, an epoch in earth's history on which humans have left their mark—the buzzword Wasteocene has been coined to point up capitalism's contaminating wastefulness.[14]

The second issue is that of *overlay*: Humans have always produced refuse, but what they left behind in earlier eras generally became compost—i.e., a beneficial element in natural life cycles. In today's world, by contrast, our waste is much more complex and persistent. It is at odds with the natural environment, a technosphere that is gradually beginning to overlay this natural world until, taken to its logical extreme, what we throw away comes to cover the entire planet.[15] In this way, waste can become a metaphor for how nature is influenced by humans, as encapsulated in the concept of the Anthropocene.[16]

And finally, there is *devalorization*: When we designate something as waste, we are declaring it to be useless and worthless. This gives rise to a strong social pressure to throw it away, to dispose of it. However, this devalorization does not simply relate to things, it can also apply to people. At the same time, it is also associated with hope: Devalorization is a condition that dissolves social hierarchies, making new constellations possible. The bankrupt banker and the single mother meet at the dump—people who, in the world of high-rises and well-lit shopping streets, would probably never form a more intimate relationship with one another.[17]

BROAD AND NARROW CONCEPTS OF WASTE

Wastage, overlay, and devalorization delineate key relationships to society that trash can be instrumental in producing. Somewhere between an analytical tool and a metaphor, waste can be used illustratively to parse universal strategies of inclusion and exclusion and make them manifest in artistic discussion. We are dealing here with a *broad* concept of waste: Trash is not just what ends up in the bin, but everything that is amenable to these strategies of inclusion and exclusion. At the risk of hyperbole, everything can be waste, and everything can become waste.

However, this can easily result in us losing sight of the actual topic. In sociological and ethnological debates in particular, there is a tendency to explain waste simply as a facet of social norms, which also means abstracting it from its concrete materiality. As if waste would dissolve into thin air if social norms were to change. A *narrow* concept of waste would thus focus on things that are explicitly accepted as garbage. This scales down the topic considerably, making it a specific aspect of our everyday lives and a discrete facet of a global environmental problem. However, from a historical perspective, for example, it helps us not to lose sight of the bigger picture. It also makes it possible to bring different disciplines into dialogue with one another, while the broad concept of waste is likely to be more or less impenetrable to engineers, for example.

There is, in the end, no need to choose one of these alternatives over the other. Rather, it is a matter of developing an awareness of the social relevance of refuse and reflecting on the factors that make waste into waste. It may be that the particular potential to be found in engaging artistically with waste lies in mediating between a narrow and a broad conception of it: Waste, in its material form, is always present, so there is no danger of reducing everything to a discourse. At the same time, the artistic proposition points beyond our everyday understanding of waste and becomes a powerful means of exploring societal strategies of devalorization and exclusion.

Roman Köster

Roman Köster is an economic and environmental historian, whose research is focused on the history of economic crises and of waste. He is an associate professor (PD) at the University of the Bundeswehr Munich. His book *Müll: Eine schmutzige Geschichte der Menschheit* (C. H. Beck, 2023) was nominated for the 2024 German Non-Fiction Prize.

Notes

1 “Gesetz über die Beseitigung von Abfällen,” *Bundesgesetzblatt* 49, June 10, 1972, 873.
2 See also Niklas Luhmann, *Ökologische Kommunikation: Kann die Gesellschaft sich auf ökologische Gefährdungen einstellen?*, 5th ed. (VS Verlag für Sozialwissenschaften, 2008), 41.
3 *Bundesgesetzblatt*, “Gesetz über die Beseitigung von Abfällen,” 873.
4 A typical example of this can be found in Bundesministerium für Umwelt, Naturschutz, nukleare Sicherheit und Verbraucherschutz, *Wertschätzen statt Wegwerfen: Konzepte und Ideen zur Abfallvermeidung* (BMUV, 2023), 5, 28, and 30.
5 Aida Bosch, *Konsum und Exklusion: Eine Kultursoziologie der Dinge* (transcript, 2010).
6 Bruno Latour, *Politics of Nature: How to Bring the Sciences into Democracy*, trans. Catherine Porter (Harvard University Press, 2004).
7 See Manfred Sommer, *Sammeln: Ein philosophischer Versuch* (Suhrkamp, 1999).
8 Scott Herring, “Collyer Curiosa: A Brief History of Hoarding,” *Criticism* 53 (2011).
9 See, for example, Marie Kondo, *The Life-Changing Magic of Tidying: A Simple, Effective Way to Banish Clutter Forever*, trans. Cathy Hirano (Vermilion, 2015).
10 On this, see Roman Köster, *Müll: Eine schmutzige Geschichte der Menschheit* (C. H. Beck, 2023), 91–102.
11 Mary Douglas, *Purity and Danger: An Analysis of Concepts of Pollution and Taboo* (1966; Routledge, 2002).
12 Violetta Simon, “Unterwäsche soll gar nicht steril sein,” *Süddeutsche Zeitung*, December 7, 2023.
13 Dietmar Rübel, “Abfall,” in *Lexikon des künstlerischen Materials: Werkstoffe der modernen Kunst von Abfall bis Zinn*, ed. Monika Wagner et al. (C. H. Beck, 2002), 13.
14 Marco Armiero, *L’era degli scarti: Cronache dal Wasteocene, la discarica globale* (Einaudi, 2021).
15 Peter K. Haff, “Technosphere,” in *Handbook of the Anthropocene: Humans Between Heritage and Future*, ed. Nathanaël Wallenhorst and Christoph Wulf (Springer, 2023).
16 On the Anthropocene, see Christophe Bonneuil and Jean-Baptiste Fressoz, *The Shock of the Anthropocene: The Earth, History and Us*, trans. David Fernbach (Verso, 2016).
17 Köster, *Müll*, 10–16.

TRANSFORMING THE CIRCUITS OF ELECTRONIC WASTE

Jennifer Gabrys

It is now common to hear warnings about the environmental impact of AI. News articles, academic research, and policy reports warn of the rising energy consumption and carbon emissions that AI contributes to, primarily through the expansion of data centers. While it is no doubt useful to raise awareness of the environmental costs of AI, these ways of framing its impact can also overlook the more extensive and uneven social and material infrastructures of computing, which are not easily captured in energy-use calculations, carbon-emissions metrics, or even life-cycle analyses. As long-standing research at the intersection of digital technology and the environment has demonstrated, there are multiple global impacts caused by computing, from mining for critical and raw minerals to pollution at electronic manufacturing sites and the recycling and disposal of electronic waste. At the same time, workers across these multiple supply chains and residents near sites of extraction, production, and disposal can experience significant harm to the environment and their health.

AI is often presented as a radical remaking of computing, but, in many ways, it is an extension of ongoing computational processes. It promises to be a digital accelerant, contributing to social, economic, and ecological tipping points. Rather than viewing AI as a radical break with previous digital developments, it should be analyzed within the context of the extensive scholarship on the environmental impacts of digital technology, which includes social and climate justice concerns. In addition to its reliance on data centers and the expansion of energy sources and grids, AI is also built on an excess of rapidly obsolescing chips, computer hardware, operating systems, and any number of new, as yet unlaunched gadgets that will become everyday wearable AI assistants.

The debris of such digital technologies often—although not exclusively—circulates from the Global North to the Global South. The workers who repurpose, repair, and recycle these devices are usually among the most disenfranchised laborers. These workers are exposed to hazardous metals, carcinogenic compounds, and unsafe working conditions as they find ways to transform the unwieldy materialities of computation. In this way, several works in the exhibition *Waste* trudge through the dense materialities of electronic devices as they travel around the world, are picked apart, repaired, or recycled, and become fossilized forms, decaying into soil and water and accreting into other ecologies. The material, social, and environmental impacts of digital technology raise questions about how to grapple with the environmental damage and injustice caused by computing devices. How could it be possible to move beyond a typical energy footprint or standards-based analysis of computing and its impacts to engage more fully with a project of digital environmental justice?

The works in *Waste: An Exhibition about the Global Routes of Rubbish* provide several possible responses to this question. In Karimah Ashadu's *Brown Goods*, we learn how African migrant workers in Hamburg, Germany, create economic opportunities by participating in the second-hand goods trade, where items such as computer monitors, television screens, and assorted appliances are exported to Africa for resale or recycling. The precarity of migration meets the refuse of computing in this unsettling study. In *TC – 2000,* Akwasi Bediako Afrane relocates seemingly remote computer discards by transporting exhibition visitors into the thick of a speculative cityscape composed of the debris of computational aspirations. While the comfort of the console may be the portal into this installation, its content asks viewers to engage

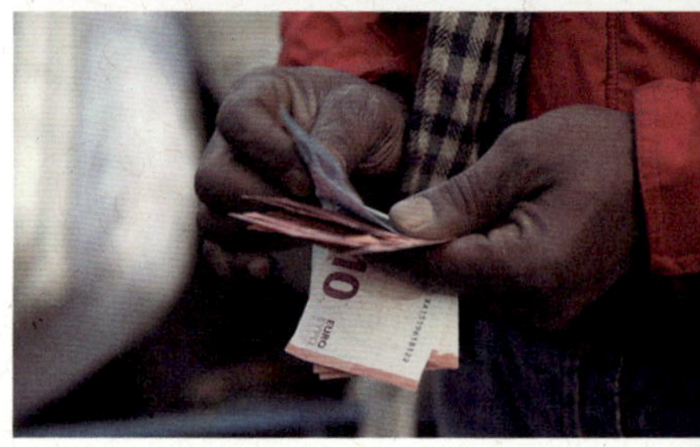

A Karimah Ashadu, Brown Goods, 2020

with the ethics of waste and to consider how leftover devices create responsibilities for electronics workers, who are charged with the tasks of reuse and recycling, hacking and transforming.

My own work on the materiality of digital technologies began over twenty years ago with a study on electronic waste. In *Digital Rubbish: A Natural History of Electronics*, I develop this processual and material approach to electronics by focusing on how e-waste is generated, from manufacturing to disposal.[1] In common with many of the works in the *Waste* exhibition, I consider how digital technologies become fossilized forms, bearing the traces of the social, economic, and material forces that contributed to their making and breaking. In this sense, *Digital Rubbish* involved a spatial-material analysis of digital technologies, and it worked through the sprawling effects of unruly technologies as they remade soil and systems, markets and consumption, time and infrastructures. I undertook this work by focusing on digital-material processes and asking: What are the material ecologies of digital technologies? How do they dematerialize and rematerialize socio-environmental relations? How might they be remade or reworked to be less damaging, more democratic, more just?

As this work and subsequent related texts discuss, the residues from digital devices include everything from discarded electronics at the end of their life cycles to resource-intensive manufacturing processes, information overload, software obsolescence, intensive energy and water use, and devices accumulating in landfills and polluting soil and water.[2] Electronic waste is one of the fastest-growing waste streams worldwide. The volumes of e-waste that are generated have risen from estimates of between 20–25 and 35–40 million tons per year in 2009 to 50 million tons per year in 2022. Exact volumes of e-waste are difficult to estimate owing to the shadowy ways in which these materials are classified and circulate, but volumes are forecast to double by 2050 to 100 million tons per year.[3] Electronic waste is hazardous, difficult to recycle at end-of-life, and often processed in harmful ways. Lead, mercury, and brominated flame retardants are just a few of the harmful chemical-material components that are used in these technologies.[4] At the same time, there is a lively group of right-to-repair movements re-making and extending electronics. Still, electronic devices are designed and manufactured in ways that can make repair difficult and even unsafe or illegal.

In addition to requiring intensive mining practices that create sacrifice zones and pollute environments, electronics generate hazardous waste during manufacture. The working conditions involved in their making and recycling are typically harmful to human health, raising significant environmental justice issues for workers and residents.[5] Early research into the electronics industry has documented impacts on workers across the electronics supply chain. Collections such as *Challenging the Chip* are important contributions at this intersection of digital technology and environmental justice, with texts from members of the Silicon Valley Toxics Coalition, a long-standing environmental justice organization in California's Bay Area that formed to address the local contamination of soil and groundwater from microchip manufacturing.[6] Many production facilities have now moved to locations outside the US and Europe; the associated residues from electronics disposal and the pollution from electronics mining and manufacturing have created geospatial inequalities and environmental injustices, typically concentrated in poorer areas of the Global South.

Despite these ongoing studies, there remains a widespread sense that digital technologies are resource-light and may even promote green alternatives by using fewer resources than analog equivalents or by enabling the ongoing monitoring of consumption activities. Although digital technologies appear immaterial, as the environmental issue of electronic waste indicates, their material effects are significant. Materiality refers not merely to the raw physicality of stuff but also to socio-ecological relations and lived environments, and to the processes by which they sediment, endure, and transform. While electronic waste demonstrates the materiality of digital media, it is emblematic too of the actuality of all that is *solid* in contrast to the apparently *virtual* movement of information. As I note in *Digital Rubbish*, electronic waste demonstrates the processes of materialization that digital media are entangled with. These processes include our contemporary material cultures of technological fascination, repetitive cycles of consumption, built-in obsolescence, poor resource use, and labor inequalities, in addition to environmental pollution.[7]

In other words, digital technologies remake worlds. In this sense, the material effects of digital technologies cannot be calculated simply through a life-cycle analysis, in which physical inputs and outputs are added up and assessed for damage to be remedied or resources to be recycled. Instead, it is necessary to attend to the relations, practices, and ways of life that are put in place and sustained through these material arrangements. *Digital Rubbish* analyzes the materiality of electronics from this perspective, discussing the multiple forms of waste they generate as evidence of the resources, labor, and imaginaries bundled into these machines. I consider the far-reaching material and cultural processes that enable the making and breaking of these technologies, which I develop into a natural history method that considers the promises and debris of computational technologies through their fossilized forms. This processual approach involves tracking the material residues of electronics, from Superfund sites contaminated with the debris of microchip manufacturing to digital markets, recycling sites, archives, and landfills, as a means to unfold and analyze the material-political processes that contribute to the making and breaking of digital technologies. This is a way not just of documenting the material ecologies and economies of these technologies but also of developing possible sites and strategies for *creative intervention and transformation*.

As the *Waste* exhibition also demonstrates, the fossils of electronic waste can be examined not just in their processes of accumulation and decay but also in the extraction and mining that precedes them, where the minerals required to power computational systems also remake planetary environments. With Ana Alenso's *Obsolete Swing*, mining processes and the sacrifice zones they produce are central to an installation that travels from mining to obsolescence, future fossils, and digital life remade. Circulating on an infinite production loop, electronics that would power computation, including AI, cycle from extraction to waste and back again. As though anticipating the floods that will submerge discarded technological devices amid the further destruction wreaked by climate change and rising sea levels, in *Cartes Mères* Hicham Berrada accelerates the material decay of metal from printed circuit boards by immersing devices in electrolytic baths. The "motherboards" of digital devices spawn crystals and resin, creating new ecologies. Beyond salvaging

B Hicham Berrada, Carte mère #19, 2020–23

useful devices from broken-down electronics, Kristof Kintera composes ruined geo-ecologies of transistors, circuit boards, copper wires, and plastic housings in the desert landscapes of *Postnaturalia*. In this work, digital hardware accretes into the ruins of an imagined energy- and information-hungry civilization, once frantically circulating and processing data, and here clogged with the silent confusion of residual keyboards and printed circuit boards.

In discussing electronic waste as an environmental issue, it is necessary to bring in the complex material cultures of digital technologies, including the apparently virtual or immaterial qualities of these technologies, the environmental health impacts and unjust working conditions that are a part of their manufacture, the digital economies that revolve around increasing rates of electronics consumerism and obsolescence, and the accumulation of e-waste and environmental fallout that comes with the decay of these technologies at the end of their life cycle.

AI is the latest sleight of hand in the attempt to erase the materiality of the gadgets and infrastructures that are central to computing. But it is important to situate AI within the ongoing digitalization of the planet, since an increasing number of gadgets are now electronic, networked, and generating data, which in turn feeds AI processes. While computational technologies are meant to contribute to greater efficiency and sustainability, and “green” and “smart” promise to provide the solution

C Krištof Kintera, Postnaturalia, 2016/17

to an overheating planet, these devices are simultaneously contributing to environmental injustices across the Global North and South, as well as within economically disadvantaged locations worldwide, creating new and expanded zones of extraction and destruction, consuming ever more energy and water while polluting soil and water and leaving behind a trail of obsolete devices for precarious workers to repurpose. With works focusing on electronic residues, the *Waste* exhibition demonstrates how geospatial inequalities and processes of materialization are remaking the planet, creating ecologies and circuits of waste that require radical transformation.

Jennifer Gabrys

Jennifer Gabrys is a professor at the University of Cambridge, where she leads the Planetary Praxis group. She is editor of the *Planetarities* series (Goldsmiths Press) and author of *Digital Rubbish* (University of Michigan Press, 2011), *Program Earth* (University of Minnesota Press, 2016), *How to Do Things with Sensors* (University of Minnesota Press, 2019), and *Citizens of Worlds* (University of Minnesota Press, 2022).

Notes

1 Jennifer Gabrys, *Digital Rubbish: A Natural History of Electronics* (University of Michigan Press, 2011).
2 Jennifer Gabrys, "Re-Thingifying the Internet of Things," in *Sustainable Media: Critical Approaches to Media and Environment*, ed. Nicole Starosielski and Janet Walker (Routledge, 2016).
3 Jennifer Gabrys, "Electronic Environmentalism: Monitoring and Making Environmental Crises," in *The Routledge Handbook of Ecomedia Studies*, ed. Alenda Chang et al. (Routledge, 2024).
4 Ruediger Kuehr and Eric Williams, eds., *Computers and the Environment: Understanding and Managing Their Impacts,* Eco-Efficiency in Industry and Science 14 (Kluwer Academic, 2003).
5 Another important early work on this topic that encompasses mining-related hazards is Elizabeth Grossman, *High Tech Trash: Digital Devices, Hidden Toxics, and Human Health* (Island Press, 2006).
6 Ted Smith et al., eds., *Challenging the Chip: Labor Rights and Environmental Justice in the Global Electronics Industry* (Temple University Press, 2006).
7 Gabrys, *Digital Rubbish*.

Illustrations

A Karimah Ashadu, Brown Goods, 2020, single-channel video installation, © Karimah Ashadu and Sadie Coles HQ, London
B Hicham Berrada, Carte mère #19, 2020–23, Courtesy of the artist, Galerie Mennour, and Wentrup Gallery; photo: Hicham Berrada, © Hicham Berrada
C Krištof Kintera, Postnaturalia, 2016/17, mixed-media installation, © Krištof Kintera; photo: Krištof Kintera

8748
L330
4.2
ppe
W30880-Q3050-A1-5
INNOVATRON
PATENT

WASTEWAYS

Amanda Boetzkes

Waste is upon us as never before. While the question of how to cope with waste has always driven the spatial and social formations of human civilizations, today its material expressions exceed our technological capacities to contain or otherwise dispose of it. Waste oozes from, chokes up, and clutters human infrastructures. What critical and aesthetic reflection can be derived from the visibility of waste, then? Museum Ostwall's *Waste* exhibition takes us into the visceral sensations and cultural imaginaries at stake in this predicament.

THE ETERNAL RETURNS OF WASTE

Waste travels the globe. It is exchanged, sorted, and transferred before it reaches zones of accumulation. One could therefore claim that waste has as much to do with the economy as it does with the environment. By the same token, the perspectives we glean from waste show us that our earthly existence is both aesthetically charged and grounded in either the economy or the environment. Our planetarity weaves the economic, environmental, and aesthetic dimensions together.

Consider Nancy Holt's design for *Sky Mound* (1985), a work that captures the critical ambitions of the Earthworks movement, which emerged in the US in the late 1960s—ambitions that, in some respects, remain unfulfilled. Holt was enlisted to design a reclamation

A Nancy Holt, Sky Mound under construction, 1991

project for the I-A Landfill in the New Jersey Meadowlands. The work itself was only partially realized, however: Holt created a pond that would drain and remediate surface water from the landfill and attract wildlife back to the wetland ecosystem, fed by the Hackensack River. Holt's full proposal, however, involved an architectonic shaping of the landfill that would include viewing areas, connected by a circuit of gravel pathways that tracked the movement of the stars and planets according to the solstices, a gunite sphere to stand in for the moon, and dramatic methane flares that would burn the gaseous runoff and be visible from airplanes overhead. Holt's visionary proposal upholds the terms of ecological perspective writ large: uniting an aerial perspective of a human dumping ground with the ordered trajectory of the celestial bodies above. More than a recycling of waste, Holt sought to recast the entire site with a consciousness of our fixedness to an axis between earth and sky.

There is no escaping waste. As sociologist Myra Hird tells us, waste has a paradoxical complexity: Solving one aspect of (any) waste crisis simply results in the introduction of newer, more complex issues further down the road.

B Nancy Holt, Bird's-eye view of the original site for Sky Mound, 1984

C Nancy Holt, Sky Mound under construction, 1991

Even when recycling technology, for instance, is improved, different challenges tend to arise—like increased toxic runoffs from new systems of recycling.[1] But art tells us that when waste returns to us, its aesthetic dimension is especially charged. Waste carries a sensibility, perspective, and critical form that is undeniably integral to the cultural zeitgeist of the late twentieth and early twenty-first centuries.

Back in the 1960s, the French New Realist artist Arman identified a commonality between the commodified appearance of art and accumulations of waste. His "accumulation" works gather together waste of all varieties—sometimes objects of a kind, like discarded electric razors, sometimes general garbage. Piling it into minimalist vitrines, Arman transforms waste from an unseen disarray into an objectified form, one that can be scrutinized with the full force of modernist aesthetic criteria. Arman directed his gesture toward the economy of modern art, which presumes to offer a freestanding and transparent presentation of form, while remaining secretly tied to the channels of exchange by which all other objects circulate. In a Marxist vein, the "secret" of the art commodity is nothing short of a degenerating interiority that pollutes the aesthetic experience from within its very formalist pretensions. Waste makes itself known as the rotten core of the economy itself. From this perspective we can have only one realization: Art is impossibly tethered to a capitalist ecology that cannot purge itself of its own waste.

D Nancy Holt, Sky Mound: Sun Viewing Area with Pond and Star Viewing Mounds, 1985

WASTEWAYS AND PARTITIONS OF THE SENSIBLE

If waste here is the concealed "truth" of the commodity (even art) as per Arman's formulation, this is so because the capitalist economy is not a freestanding system; it is a priori an ecology. Which is to say that to the extent that commodities travel global economic pathways from their sites of production to their sites of consumption, they bear the traces of their environmental effects. The visibility of waste since the end of the twentieth century shows us the dysfunction of the most functional human technologies: obsolete circuit boards, plastic computer hardware, copper wire, e-waste, all interspersed with organic detritus. Contemporary waste is cultural as never before.

Recycling initiatives are only a partial remedy for the scale and quality of ecological decline. Nevertheless, they signal to us an emergent ecological consciousness. Carbon emissions and rising global temperatures remain untouched by the breakdown and recirculation of objects. Yet, the recent concern with waste as a problematic that demands intellectual, scientific, and creative innovation—the very reconstruction of a lifeworld of ecologically robust relations—is corroborated through contemporary art's new partitions of the sensible.
The suggestion that sensibility is partitioned is a slightly different reflection from Jacques Rancière's concept of the "distribution of the sensible." Ecological culture

does not develop out of the politics of aesthetics so much as it issues from the material substructures that govern labor, resources, knowledge, and bodies, in terms not only of their function but, just as importantly, of their dysfunction as well. Which is to say that ecological aesthetics exist as a kind of recoil of colonial capitalism. As such, art is oriented toward the imperatives to feel and know the planet in the wake of its exploitation and abuse. As philosopher Michael Marder argues, the schema at work in capitalism's current energy imaginary can be characterized by the desire to liquidate potential energy from the static, inward-dwelling energy of all beings and things.[2] Under the spell of these "energy dreams," the conceptual and spatial framing of the very notion of nature has collapsed. Yet, Marder's characterization of the cultural imaginary comes with an ethical imperative as well: to find the courage to admit to our unity with the violated earth, one that has never been conceived of before. This unity with the earth, which has been "fracked, disemboweled, and filled with our garbage," stems from a unity of love and knowledge and an effort to pass from that love to ethical life.[3]

The admission of our unity with the transgressed earth involves taking hold of the very queerness of human bodies, not only in their ontological estrangement from the natural order of things but in their convergence with toxic conditions of planetary wasting. Toxins, as cultural theorist Mel Y. Chen explains, threaten to alter our realms of comfort and current socio-cultural norms. Chen suggests that we must think *with* toxicity to consider the connections between the affects related to toxicity and the sociality of queerness.[4] They propose this as a remedy for the capitalist values that inform the standardized fear of toxicity, or desire for immunity, precisely because the affected bodies are considered invalid in existential terms. Given the rapidly growing use of toxin-procuring materials like plastic and lead, there is a certain "persistent allure" to re-signifying the ontological connotation of toxicity by crediting queer bodies for already living with the attributes most associated with toxins: non-normative traits of identity (like gender, or sexuality), and non-biological reproductivity.[5]

Chen assesses the racialized fear associated with lead toxicity, constructing racialized bodies as toxins that have a kinship with the chemical pollutants produced by lead. Toxic assets become the materialized extensions of toxic bodies.[6] If toxins literally, and ontologically, threaten to alter our interior composition, then everything becomes toxic. Queer bodies are often treated as toxic, based on the segregation of queer bodies and the biopolitical privileging of heteronormative bodies since the nineteenth century at the latest. Of course, since toxicity spans the boundaries of "life" and "nonlife," in some sense toxicity releases the capacities of life from affected bodies. Ultimately, Chen invites a sympathetic attitude to intoxication, so that we might function *with* toxicity.[7]

E Mika Rottenberg, Squeeze, 2010

In this vein, we can examine Mika's Rottenberg's video *Squeeze* (2010), a curious representation of the partitioning of bodies, sensations, female labor, and waste in disparate but interlinked domains of the global economy. *Squeeze*, which was not part of the *Waste* exhibition, shows us the inverse perspective of Arman's accumulation sculptures. Rather than introjecting waste into the commodity form of art, Rothenberg takes us into a labyrinth of commodity production. BIPOC women work as field hands on a lettuce farm in Arizona or harvest sap from trees in a latex forest in India. Asian women work in isolated compartments, periodically massaging oils into the hands and arms of the laborers. A Black woman sits on a round turnstile, seemingly providing the weight for the various mechanisms that squeeze out living essences. A white woman oversees the operations of the system from an underground compartment, idly eating a sandwich. A tongue protrudes from a hole in the wall, and a glass of unidentified yellow fluid enters the space from a pulley system. The women are radically separated from each other, present only as arms, tongues, eyes, exposed buttocks, faces, and hands that issue secretions, dirt, saliva, and skin debris or, alternatively, may require tending: sprayed water, massage oil, an ice bath. The by-products of the women's laboring body parts are themselves integral to the system. Marx characterizes this siphoning of bodily energy into a systemic operation as "living labour." The laborer invests energies into the

product, yet the laborer's body and its instruments are at once consumed by the system of product, "bathed in the fire of labour."[8] Rothenberg's vision of global labor is one in which the worker is not so much alienated from the commodity as absorbed into a systemic predicament: bodies, their labors, products, and wastes, brought together in a queer intimacy.

In *Squeeze*, there is an uncanny vitality to the partitions that bind, separate, and contain the working women. The boundaries between them are oddly textured in their physical contiguity with the bodies. Sweat mingles with perfumed oil, sprays of water with saliva, flesh with strips of latex: The temperature rises, and the energies of the labor system generate a miasma. This is what Chen would call the "animacy" of toxicity, the nonreproductive but

nevertheless indomitable presence of agents that persist in states between life and death, organic and inorganic, subjects and objects. This animacy is relational, non-normative, and ultimately queer in its interruption of hetero-normative environmental ideology.[9] The animacy of toxic wastes might therefore be considered for their potential to renew social relations even as they degrade environments and endanger the living.

THE FUTURITY OF WASTE

Waste becomes the occasion of social renewal as its animacy enters the domain of aesthetic inquiry. Tejal Shah's multi-channel video installation *Between the Waves* knots queer existence together with the toxic animacy of environments. The outcome is a cosmology galvanized by the encounters between humans and mythic beings from a forgotten civilization (with unicorn horns on their heads). Sexual communion between the human and mythic females takes place against the backdrop of a thickly polluted urban atmosphere. Shah depicts a coalescence of energies and bodies that embrace the postapocalyptic environment while at the same time resetting it as the site of a re-origination of the world. The erotic probing of bodies seemingly occurs at the beginning and end of the world, a planet that is at once a queer utopia and an ecological dystopia.

Where Shah cultivates queer intimacy and toxic animacy coextensively, a more alien futurity appears in Francois Knoetze's four-part film series *Core Dump* (2018–19). Here, human life appears as the recoil of technology, as memory bytes recovered from e-waste, as though the human world has passed and is already nothing more than an accumulation of fleeting images. The human, this archive of the future shows us, is destined to collide with the technosphere. A narrative emerges in which the Silicon Valley billionaires, in their zeal to become immortal, have built underground bunkers to continue their project of designing rockets to colonize Mars, while they ward off climate refugees aboveground. Protagonists and antagonists rise up from the agon of technology and planetary life; the landfill becomes a transhuman reality. Waste is not condemned but is, rather, the marker horizon on which humanity has left its mark. The human is an afterthought of technology, its visibility mediated by the screens, microchips, and circuit boards that bear the traces of its image.

G Francois Knoetze, Core Dump – Kinshasa, 2018
H Francois Knoetze, Core Dump – Shenzhen, 2018
I Francois Knoetze, Core Dump – New York, 2019
J Francois Knoetze, Core Dump – Dakar, 2018

Waste invites us to reflect as much on the futurity of waste as on its aesthetic history. The movement between history and futurity has a bidirectional trajectory that speaks to possibilities of social ingenuity, bodily intimacy, and the environmental imaginary as well as to the troubled past of exploitation, extraction, and toxic dumping.
It shows us the synthesis of exchange and accumulation; animacy and inanimacy; utopia and dystopia; humanism and the inhumane at play in charting the ways of contemporary waste.

Amanda Boetzkes

Amanda Boetzkes is Research Leadership Chair and Professor of Contemporary Art History and Theory at the University of Guelph. She is the author of *Plastic Capitalism: Contemporary Art and the Drive to Waste* (MIT Press, 2019) and *The Ethics of Earth Art* (University of Minnesota Press, 2010) and co-editor of *Art's Realism in the Post-Truth Era* (Edinburgh University Press, 2024), *Artworks for Jellyfish and Other Others* (Noxious Sector Press, 2022), and *Heidegger and the Work of Art History* (Routledge, 2014).

Notes

1 Myra Hird, *Waste: The Basics* (Routledge Press, 2025).
2 Michael Marder, Energy Dreams: *Of Actuality* (Columbia University Press, 2017), 56.
3 Michael Marder, "For the Earth That Has Never Been," in "Terra, Natura, Materia," *Stasis* 8, no. 1 (2020): 74.
4 Mel Y. Chen, "Toxic Animacies, Inanimate Affections," *GLQ: A Journal of Lesbian and Gay Studies* 17, nos. 2–3 (2011): 265.
5 Chen, "Toxic Animacies," 266.
6 Chen, "Toxic Animacies," 270.
7 Chen, "Toxic Animacies," 273.
8 Karl Marx, *Capital: A Critique of Political Economy*, ed. Frederick Engels, trans. Samuel Moore and Edward Aveling (The Modern Library, 1906), 204.
9 Chen, "Toxic Animacies," 280.

Illustrations

A,C Nancy Holt, Sky Mound under construction, 1991, © Holt/Smithson Foundation, licensed by Artists Rights Society, New York

B Nancy Holt, Bird's-eye view of the original site for Sky Mound, 1984, © Holt/Smithson Foundation, licensed by Artists Rights Society, New York

D Nancy Holt, Sky Mound: Sun Viewing Area with Pond and Star Viewing Mounds, 1985, graphite on paper, © Holt/Smithson Foundation, licensed by Artists Rights Society, New York

E Mika Rottenberg, Squeeze, 2010, single-channel video installation with sound and digital C-print (film stills), © Mika Rottenberg, Courtesy of the artist and Hauser & Wirth

F Tejal Shah, Between the Waves, 2012, five-channel video installation (film still), © Tejal Shah and Project 88

G Francois Knoetze, Core Dump – Kinshasa, 2018, single-channel video installation (film still), © Francois Knoetze

H Francois Knoetze, Core Dump – Shenzhen, 2018, single-channel video installation (film still), © Francois Knoetze

I Francois Knoetze, Core Dump – New York, 2019, single-channel video installation (film still), © Francois Knoetze

J Francois Knoetze, Core Dump – Dakar, 2018, single-channel video installation (film still), © Francois Knoetze

MANUFACTURED LANDSCAPES*:

ON WASTE AS INFORMATION

Susi Gutsche

WAYS OF SEEING: IN THE BEGINNING WAS THE MATERIAL, AND THE MATERIAL WAS TRASH

Waste is not the end but the beginning, the information. What we discard tells us more about who we are than what we keep does. Waste is culture. Since the beginning of human activity, waste has been connected to cultural development. Waste accumulates as cultural artifacts, and art is a material practice.

Museums have always been full of trash. Cultural developments over thousands of years are trash, in a totally neutral sense. From human remains to everyday objects, materials bear witness to cultural, economic, ecological, and political histories. Once broken and no longer suitable for their original use, ancient ceramic fragments were engraved with everyday writings and thereby reused and recycled. In this way, objects transformed their functions over time. These ceramic shards served democracy: In ancient Athens, citizens scratched names into pottery fragments called *ostraka* to vote on banishment.[1] It shows that waste has always carried democratic values. But now? Waste in all its forms, solid, liquid, and gaseous, is becoming our planetary breaking point, eroding democracy. Trash remains an instrument of (colonial) power, yet its very omnipresence transforms it into a voice that cannot be silenced and

*The title references the documentary *Manufactured Landscapes* (2006), featuring Edward Burtynsky's industrial landscape photography, directed by Jennifer Baichwal.

thus speaks for itself. Waste is a carrier of information, a carrier of culture, a reminder of the past.[2]

The Great Acceleration:[3] Today, we no longer vote with engraved shards. While waste grows inexorably, it accumulates both metaphorically and materially, with plastics at the center of it. Multilateral endeavors collapse, global efforts to reduce (plastic) production and therefore pollution have largely failed. Petrostates rise, feeding global trade markets, and accelerate the growth of (new) waste colonies. Out of mind, out of sight, out of territory becomes the mantra on every level: governmental and societal, ecological, and of course economical.

Not in my backyard!

Using Erwin Panofsky's method, we can read waste through three levels: typological, iconographical, and iconological.[4] Borrowing from art history, we learn to see in layers. First, we observe only what is there: color, shape, texture. A line. A brown cylinder. A mountain of fragments. No interpretation, just seeing. Then, we begin to name what we recognize: ceramics, plastic, a Roman refuse heap, a contemporary landfill. Finally, cultural knowledge enters and allows synthesis: What does this reveal about the world that made it? How do we interpret these accumulations? To understand waste, we must learn to read it as we read art: seeing its pure form, recognizing its cultural meanings, synthesizing what it reveals about the systems, values, and futures that produced it.

In summary, waste operates simultaneously as object, as symbol, and as evidence of ideology. In the beginning was the material. And the material was already trash.

FORM FOLLOWS FUNCTION: THE CHOREOGRAPHY OF DISPOSAL, SEEING PURE FORM

Now let's imagine we are looking at a plastic mountain and a ceramic mountain.

Ceramics: clay ▶ fired ▶ broken ▶ form a mountain
Plastic: fossil fuels extracted ▶ processed
▶ manufactured ▶ consumed ▶ discarded
▶ toxic landfilled mountain

The traces of human activity appear in the ambiguous state of being visible and invisible at the same time. From above, these mountains resemble natural geological formations, sedimentary deposits. When we look closer, every fragment is a decision, a gesture of disposal. At the level of pure phenomenology, the pre-iconographical level, we observe form without understanding. Both hills reach the same height. Both create topographies through accumulation. Layer upon layer, year upon year.

Stratification, compression, the physics of disposal. They are manufactured landscapes that could pass for natural formations, yet both are accumulated geographies of human activity, telling stories of what we consumed, how we lived, what we discarded.

Forms are never innocent. The ceramic mountain becomes Monte Testaccio consisting of millions of amphora fragments documenting Rome's Mediterranean trade.[5] The plastic mountain becomes evidence of contemporary consumption, planned obsolescence, and borders that waste crosses while people cannot.

Waste moves between archaeology and real-time assessment, between ancient hills and contemporary landfills. Waste becomes a time machine of history, a walkable excavation. It exists in the stratigraphy of our waste age. Visualized movements become choreographed geographies. Waste migrates, flows, forms landscapes. Trajectories and routes trace across continents. Never static, always in motion.

We look at objects the way archaeologists examine spongy structures in soil sections: as maps, as digital cuts, as collage and assemblage. When we bring cultural knowledge to these forms, we begin to read them differently.

Form follows function, but function also follows necropolitics:[6] Who disposes, who is exploited, who dies under the weight of accumulated materials in soil, water, and air? Here we reach the level of iconology, of synthesis:

These landscapes reveal not just what we discarded but why we discarded it, where we sent it, and who bears the burden of our disposal. Who is impacted directly by our "Western" toxicity?

Contemporary waste colonies exemplify necropolitics through the creation of what Achille Mbembe calls, in the conclusion to his correspondingly titled text, "*death-worlds*, new and unique forms of social existence in which vast populations are subjected to conditions of life conferring upon them the status of *living dead*."[7] Populations are abandoned to "let live and make die" amid toxic e-waste and plastic mountains. The Global North's performative environmental rhetoric empties waste colonialism's abnormal reality of meaning through what Marina Gržinić describes as "a repetitive performative mechanism that functions as a process of extracting meaning, in the sense of voiding or emptying."[8] Waste disposal follows racialized pathways where whole populations bear the toxic burden. This necropolitical choreography reverses colonial extraction routes: Instead of resources flowing out, toxicity flows in, as workers sort through toxic materials in places like Ghana's Agbogbloshie, the resting place of much of Europe's e-waste.

We ask not just what a phenomenon is or what it represents, but what it reveals. We synthesize. The Romans built Monte Testaccio because their empire consumed at scale. Trade required standardized containers.

The mountain stands as evidence of logistics, economics, technology, and imperial reach. Waste is transported globally, not from lack of alternatives but because our economic system externalizes disposal costs. To keep costs low and to get waste out of sight. But this only shifts the problem. This is not an accident. It is built into the system. It reveals what we value and externalize, who we protect and sacrifice, what futures we imagine and what ruins we leave.

These mountains are geological artifacts that rise into the environment, appearing to flow naturally, binding themselves into geographical logics. This visual land-grabbing leads to territorial appropriation.

Waste is not a closed system; it is a system that leaks, drips, and emits gases. But structurally and from the outside, all we see is a mountain.

"The owls are not what they seem" suggests that these mountains and elevations are not what they appear to be.[9] The hills symbolize environmental neglect, though this is often not visible at first sight.

Form follows function, but function follows ideology and industries.

A Jan Henderikse, Untitled, ca. 1960

TRASH PEOPLE IN A TRASH WORLD

Does art need to be beautiful? What does beauty mean? Aesthetics is something inherently subjective and always in transformation. Yet there exists a beauty in trash, a familiar aesthetic recognizable to all humans. Not despite waste, but through it—or perhaps *both* despite *and* through it. This offers a new perception of reality, a beauty in collapse and disaster. Art for art's sake at its purest expression. Waste moves between environmental and ethical catastrophes toward a new form of extinction aesthetics.

Many artists work with the iconographic traces of the discarded, anti-abstract. Waste itself becomes iconology: ceramic trash alongside plastic and other trash, both seen as aesthetically equivalent, both historically loaded. Waste in art and design is nothing new. Notably since the 1960s, artists have critically engaged with what society considers no longer usable.[10] From Arman Fernandez's accumulations, and César Baldaccini's compressed cars to Francis Alÿs's participatory projects, from sociologists to design thinkers like Victor Papanek, practitioners across disciplines have explored discarded materials over the past decades. The focus is on the connection between art and life. They established an iconographic vocabulary:

B César Baldaccini, Untitled, 1959

C Francis Alÿs in collaboration with Julien Devaux, Barrenderos (Sweepers), 2004

the assemblage, the accumulation, the compression, the transformation. We recognize these visual languages and can trace them to contemporary practice.

But the question goes deeper. What if trash is not just metaphor but an actual archive? The participative works of Francis Alÿs—particularly, *Barrenderos/Sweepers*—document waste economies and the lives of waste workers, while critiquing consumer culture.[11] These works teach us to move from seeing waste as a symbol to understanding it as evidence, as testimony, as the material residue of economic and political systems, by collaboratively building a waste barricade. Contemporary practice has its roots in the Fluxus manifesto to "PROMOTE A REVOLUTIONARY FLOOD AND TIDE IN ART.

Promote living art, anti-art, promote NON ART REALITY to be fully grasped by all peoples, not only critics, dilettantes and professionals."[12] While also acknowledging a degree of divergence, the core elements of process over product, everyday actions as art (pushing ice blocks, moving sand dunes), participatory elements (recruiting volunteers for actions), and dissolving art/life boundaries can all be connected to the Fluxus principles.

No separation exists between material and meaning, between everyday life and art, between object and participation. Artworks become environments to be entered: familiar worlds, realities constructed from found objects, data, materials, plastic fragments. The methodology moves through all three levels: We see the plastic fragment (form), we recognize it as a ballot (iconography), we understand it as a continuous material practice across millennia (iconology). No waste ever disappears in a vanishing machine. A toxic trace always remains and moves. It is unstoppable.

PATHOLOGY OF (WASTE) OBJECTS: BODIES IN FRAGMENTS AND HEALTH AS INFORMATION

Waste is fundamentally a question of health. Artifacts from ancient Vienna (Vindobona), from the Roman latrines and sewers, reveal fleas, lice, and insect larvae.[13]

Historical traces of disease leave microscopic evidence behind. Infrastructure has always distributed health and danger unequally. Historical traces of disease from ceramic chards correspond with today's microplastic analyses—both document invisible health risks in everyday objects. Ancient ceramic fragments become diagnostic instruments for ancient urban health. Contemporary plastic fragments sample environmental toxins. Ceramics become bio-sensors, each fragment a data point. Waste is information about our bodies, our cities, our future.

Formally, we see microscopic images, organic traces, material residues. Iconographically, we recognize parasites as markers of sanitation systems, microplastics as markers of petrochemical ubiquity. At the iconological level, we understand that both reveal which bodies societies deem expendable, how infrastructure distributes risk, and what "acceptable" contamination means across different historical moments. Consider our future, with "+1.5 Degrees: Malnutrition, malaria, and other diseases continue to increase."[14] Fragments and particles of our technocapitalist reality create a totalitarian impact. Decomposed materials flow as particles in wind and air, through respiration, everywhere. The extractive age simultaneously displaces us while ushering in a new era for the environment and all of Earth's inhabitants. These substances flow within us. Bodies now contain partially decomposed pollutants. Waste is becoming part of us. Ibrahim Mahama's work *Zilijifa* confronts weight on bodies with colonial pasts.[15] We see how the spine deforms over time from carrying heavy loads and the ongoing remains and weight of British colonial rule. The wounded world's aesthetics remain both invisible and visible simultaneously. This raises critical questions: Who lives on unpolluted land? Who recovers resources? Who bears the burden? Waste becomes visualized data, bacterial growth patterns, geological history read through stratigraphy. Environmental justice intersects with human rights, exposing social inequalities through the geography of contamination.

CONSTRUCT A WORLD: FUTURE ARCHAEOLOGY

In our environments, we experience how today's decisions become tomorrow's excavation layers. Visual patterns of waste distribution become archaeological maps. Digital tracking lines become settlement plans, visualizing contemporary data. We again experience the nature of waste reading: We see the stratification (form); we recognize the objects and their origins (iconography); we synthesize understanding of the economic and ecological systems that produced these layers (iconology).

We can synthesize the “plastiglomerate” as the waste object displaying “entanglement of nature and culture” in its purest form:

The plastic and beach detritus had been combined into a single substance by bonfires. Human action on the beach had created what Corcoran and Jazvac named "plastiglomerate," a sand-and-plastic conglomerate. Molten plastic had also in-filled many of the vesicles in the volcanic rock, becoming part of the land that would eventually be eroded back into sand.

The term "plastiglomerate" refers most specifically to "an indurated, multi-composite material made hard by agglutination of rock and molten plastic."16

Plastic waste presents an unknown temporality. We cannot accurately predict how microplastics will decompose in the wild or what eternal chemicals and transformations await us. As part of Earth's geology, these materials will certainly affect all planetary creatures, remaining constantly in motion through air, wind, weather, transportation, through various infrastructures and industries, and through both human and nonhuman bodies. Microplastic fragments will persist forever as records of Earth's memory, visible data, and evidence.

SCROLLING AND STROLLING THROUGH RUINS: A CODA ON CONTEMPORARY READING

We live in a world where we can scroll through feeds on Instagram, where advertisement and the fetish of new, smooth, and shiny commodities is one swipe and a digital window away from news about natural disasters. Pollution and fast fashion purchases exist a fingertip away from each other. Yet the material reality of waste cannot be wiped away with a finger. It accumulates. It persists. It speaks. Spinning worlds. This juxtaposition reveals itself in layers too. We see bright images and smooth surfaces and recognize advertising rhetoric, disaster photo-journalism, the grammar of social media. We understand that this interface design flattens information, where extinction, pollution, war, murder, and fashion share the same visual space, the same gesture of dismissal. The scroll reveals cognitive dissonance. We know about waste's consequences while participating in its production, both as witness and perpetrator. Waste is an archive, a mirror. What futures are we manufacturing?

IN THE END WE HAVE LEARNED HOW TO READ WHAT WE DISCARD

Art practices can become artificial connectors, mediators and communicators between the outside world and our

D TRACEWASTE, 2022, exhibition view and interface detail

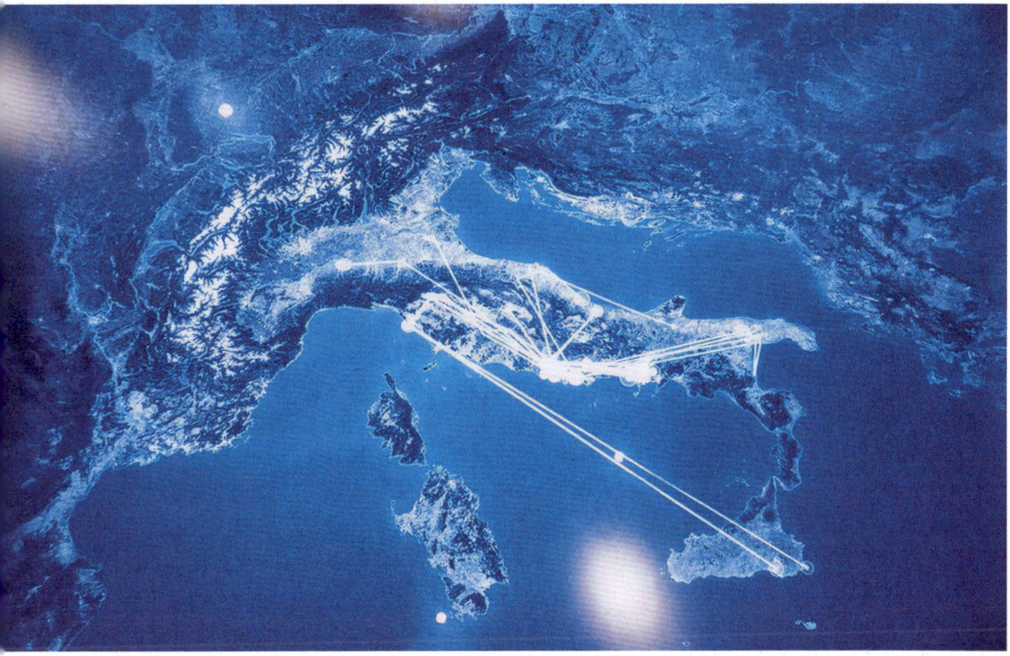

interior understanding, through materials that exist both outside and within.

The project *TRACEWASTE*, which I have been working on together with Dimitrije Andrijevic, Sebastian Scholz, and Max Pellert since 2022, exemplifies this methodology of reading waste as information, tracking discarded objects through geolocation to reveal the hidden choreography of disposal across borders.[17] By monitoring plastic and textile movements from a citizen's perspective, the project transforms invisible waste routes into visible data, creating an audiovisual installation that functions as an island of accumulated evidence. The interactive interface allows viewers to trace garbage trajectories with their own hands, while electromagnetic emissions from tracking devices create a soundscape. This work demonstrates how waste becomes both archive and oracle, not just documenting where objects go but revealing systems that determine their paths. In the beginning was the material. In the end is the material. Between beginning and end: us, making, discarding, making again.

The question is not whether we produce waste. The question is, What stories will our waste tell about us? What will emerge from our ruins?

Every pottery shard, every plastic fragment, every compressed car operates simultaneously as form, as symbol, as evidence. Learning to read waste at all three levels, phenomenological, iconographic, iconological, means understanding that material culture is never innocent, never merely functional, never outside of history. Our epistemologies will transform alongside our self-created histories. Questions of agency and spatial politics within aesthetic transformation will become our survival strategies, hopefully catalyzing positive developments and activist futures in a world full of turning points.

Perhaps we can speculate optimistically: Evolution might accelerate, and insects, bacteria, and fungi will develop to metabolize plastics. Perhaps we humans will become hybrid creatures resistant to toxins, digesting plastics, transforming our own world and the system we created. Ultimately, the transformation of nature will inevitably transform us. Renewal out of ruins. If we are living in the era of trash, what will our future look like?

Susi Gutsche

Susi Gutsche is a research-driven artist with an interdisciplinary background in Social Design – Arts as Urban Innovation, the humanities, and lab work. In her artistic practice, she is primarily interested in the intersection between art and everyday life as well as in routines and their aesthetic potential. Her focus here is on social and ecological processes. In 2026, she won the START scholarship sponsored by Austria's Federal Ministry for Housing, Arts, Culture, Media, and Sport.

Notes

1 See Wikipedia, "Ostracism," last modified December 10, 2025, 07:09 (UTC), https://en.wikipedia.org/wiki/Ostracism.
2 For further readings on urban legibility, see Dietmar Offenhuber, *Waste Is Information: Infrastructure Legibility and Governance* (The MIT Press, 2017).
3 "The Great Acceleration," *Future Earth*, January 16, 2015, https://futureearth.org/2015/01/16/the-great-acceleration/.
4 Erwin Panofsky, "Ikonographie und Ikonologie," in *Bildende Kunst als Zeichensystem*, ed. Ekkehard Kaemmerling, vol. 1, *Ikonographie und Ikonologie: Theorien – Entwicklung – Probleme* (DuMont 1994). The text describes the three-level methodology: pre-iconographical (pure form), iconographical (conventional meaning), iconological (intrinsic meaning / cultural synthesis)
5 Monte Testaccio is the largest known deposit of ancient refuse, an artificial hill of Roman amphorae.
6 Achille Mbembe, "Necropolitics," *Public Culture* 15, no. 1 (2003).
7 Mbembe, "Necropolitics," 40.
8 Marina Gržinić, "Biopolitics and Necropolitics in Relation to the Lacanian Four Discourses," paper presented at the symposium Art & Research – Shared Methodologies: Politics and Translation, Barcelona, Spain, September 6–7, 2012, n.p.
9 See David Lynch's *Twin Peaks*, season 2, episode 1.
10 From the 1960s onwards, there was a general explosion across various disciplines in critically engaging with waste and (post)consumer materials.
11 Francis Alÿs, *Barrenderos / Sweepers* (2004), video, 6:53, https://francisalys.com/barrenderos/.
12 George Maciunas, *Fluxus Manifesto* (1963), MoMA, accessed November 7, 2025, https://www.moma.org/collection/works/127947.
13 "Was haben die Römer je für uns getan? Zum Beispiel Parasiten verbreitet," *Der Standard*, January 10, 2016, https://www.derstandard.at/story/2000028702140/was-haben-die-roemer-je-fuer-uns-getan-zum-beispiel.
14 Natasha Myers, "Edenic Apocalypse: Singapore's End-of-Time Botanical Tourism," in *Art in the Anthropocene: Encounters Among Aesthetics, Politics, Environments and Epistemologies*, ed. Heather Davis and Etienne Turpin (Open Humanities Press, 2015), 39.
15 "Ibrahim Mahama: *Zilijifa*, 9.7.–2.11.2025," Kunsthalle Wien, accessed November 7, 2025, https://kunsthallewien.at/en/exhibition/ibrahim-mahama-zilijifa.
16 Kirsty Robertson, "Plastiglomerate," *e-flux journal* 78 (December 2016), https://www.e-flux.com/journal/78/82878/plastiglomerate/.
17 See www.tracewaste.eu.

Illustrations

A Jan Henderikse, Untitled, ca. 1960, Plastic box with screws, plastic figures, axles from toy cars, bottle corks, thumbtacks, shards, light bulb, plastic springs; Collection Museum Ostwall at the Dortmunder U

B Francis Alÿs in collaboration with Julien Devaux, Barrenderos (Sweepers), 2004, single-channel video (film stills); © Francis Alÿs, Courtesy the artist and David Zwirner

C César Baldaccini, Untitled, 1959, Painted automotive metal parts, Collection Museum Ostwall at the Dortmunder U © César Baldaccini; photo: Jürgen Spiler, Dortmund

D TRACEWASTE, 2022, exhibition view and interface detail, MAXXI Rome, REWILD exhibition of S+T+ARTS Repairing the Present, © Susi Gutsche; photo: Musacchio, Ianniello, Pasqualini & Fucilla, Courtesy of Fondazione MAXXI

BlackBerry
del
alt
space
sym

THE WONDERS OF WASTE:

THE VALUE OF AFROFUTURISM'S TRASH AESTHETICS

Nedine Moonsamy

Late-stage capitalism perpetuated the fantasy of excessive, unbridled consumption by turning it into a competitive sport. At the tail end of this dream lies the insurmountable problem of waste. As developed countries in the Global North grapple with the consequences of rabid consumerism, they must struggle with the growing challenges of waste disposal and recycling schemes. Proving somewhat reluctant to deal with the challenges of waste within their own borders, many countries exploit the poorly regulated flow of waste management into the Global South. Now, horrifying images circle the globe as dumps in Africa, Asia, and the Indian subcontinent grow into mountains that irrevocably alter these landscapes, ecosystems, and their various inhabitants in unpredictable ways. Yet according to critics, the fact that the Global South has been identified as a key recipient of the international waste problem mirrors historical patterns, as these spaces have always served as material *and* discursive dumping grounds for the West.

In her book *Histories of Dirt: Media and Urban Life in Colonial and Postcolonial Lagos* (2020), Stephanie Newell explores how colonial ideas on waste management and sanitation in Africa have demoralized the entire continent by using the language of dirt and waste as imperial constructs that legitimized the otherness of Africans. By assigning moral and aesthetic values to waste and dirt, those associated with it came to be seen as useless, backward, and undesirable. Hence, waste becomes a language through which "people are transformed into dirt."[1] The colonial architecture of Being instituted a racial schema where Black and African identity represented both the excessiveness and the nothingness of waste. Representations of African life were thus either over-determined or invisibilized but never granted the easy assurances of simply Being Human. In more recent times, little has changed, as it is "the dumpsite [that now] absorbs the people who enter it into a posthuman black hole."[2] The influx of waste in Africa thus reproduces, if not relies on, the coloniality of Being, sucking African existence into the gravitational force of late-stage capitalism. Many African artists have grappled with this history of dehumanizing representation, often by seeking to restore the dignity and sovereignty of African life and subjects. Afrofuturism, however, does not seek to escape from historical accusations of otherness and abjection but instead leans into them. As we see in contemporary Afrofuturism, the association with waste becomes a complex, conceptual language for considering alternative forms of identity making that belie the seductions of wanting to be human at all.

Given that the slave trade instituted states of extreme disembodiment and dislocation, Afrofuturism often uses tropes of alienation for the artistic expression of Black and African identity in popular culture. It serves to acknowledge how systemic and historical dehumanization has rendered the world post-apocalyptic at the outset, meaning that any journey toward creation becomes an unceasing reckoning with the rubble of white epistemology and ontology. Yet rather than seeking out anodyne ideologies, Afrofuturism rejects "Western concepts of subjecthood vested in a rhetoric of wholeness by forging not only a *felt* understanding of history but also an enriched self, born out of fragmentation."[3] By not seeking out recuperative strategies, Afrofuturism is drawn to the wasteland as a site for the emergence of new futures and identities for those who are otherwise denied human status.

Artists like Wanuri Kahiu (Kenya), Fabrice Monteiro (Senegal), Cyrus Kabiru (Kenya), Olalekan Jeyifous (Nigeria), Baloji (DRC) and Akwasi Bediako Afrane (Ghana), whose work is on exhibition in *Waste: An Exhibition About the Global Routes of Rubbish*, have dialed into a "trash aesthetic" by making an Afrofuturistic spectacle of waste.[4] While "dirt has been a potent category in Eurocentric representations of Africa for more than a century and remains a source of such fascination for Western publics," making it sometimes risky to reiterate as art, these creators display a cheeky audacity in their elucidation of waste.[5] While in the West, the realities of waste are ignored and made invisible through its exportation to the Global South, these artists make a performative display of the ongoing flow of oppression by putting waste on ironic display as an aesthetic form.

In his self-directed music video / short film, *Zombies*, for example, the Congolese Belgian rapper Baloji stands on a landfill in his birthplace, Lubumbashi, which, he states, is "back where it all started."[6] This statement is not a reclamation of the space, as the overwhelming presence of waste makes it apparent how consumer culture has had an irrevocable impact on the environment. Baloji stretches this sense of repugnance through visual hyperbole.

By confronting viewers with these images, he exposes Africa as a wasteland of Western design: a clear indication of the dirty underbelly of its consumerism and the role that Africa is expected to play in it. Afrofuturism thus confronts the ever-growing waste problem and uses this

A Baloji, Zombies, 2019

critique to think beyond our current global solutions of waste management.

Invented in the 1970s, and used mainly in the EU, the waste hierarchy is a socio-material response to waste management. By suggesting different options for the disassembly and reassembly of waste, it ranks "the desirability of different waste management approaches according to their environmental impact."[7] Typically iterated through the "Reduce, Reuse, Recycle" slogan, the more protracted categories are ranked as prevention, reuse, recycling, incineration, and if all else fails, landfill disposal. Hence, the vivid illustrations of the mess and materiality of consumerist technology are central to Afrofuturism as it challenges Western denialism about the effects of waste production and elimination, since most waste ends up in landfills—an indication of the waste hierarchy's collapse. This is also made clear in Akwasi Afrane's previous exhibition, *TRONS'R'US* (2023), where the mystifying aura of technology is shattered by what he describes as a "white box" process. As he explains, the idea of the "black box" is representative of how Western technology depends on people's inability to pry open its secrets. Modern technology relies on the glorification of sleek surfaces that deliberately obscure the underlying construction and the consequences of a device's design. By contrast, Afrane deconstructs these objects, taking us from product to process and highlighting the visual excess and tactile qualities of their material form.

By foregrounding waste, these artists prove that "the concept of complete recycling is a bit of a myth today. In this sense, recycling is never complete: it always generates even more waste."[8] E-waste is often shipped off to the Global South with the beneficent idea that "one man's trash is another man's treasure." Given the high turnover of electronics in the Global North, many assume that receiving containers of e-waste is a boon for the informal recycling market in Africa. In reality, much of the waste remains as such, and digging into landfills becomes a harrowing—if not humiliating—way to eke out a living for locals. In response to this sobering reality, Afrofuturism undercuts the optimistic tone and efficient claims of the waste hierarchy, thereby illuminating the void where it is necessary to reimagine alternative futures through the inclusion of waste.

In the *Shanty Mega-Structures* series by Olalekan Jeyifous, future Lagos is created by splicing and combining its current infrastructure in varying states of development. These images completely confound Western approaches to urban futurism precisely because they reject an aesthetic of sleekness. Typically, in Western futurism, images of pristine, shiny facades and tall, sharply defined buildings convey the aspirational achievements of Western modernity and capitalism by pushing waste—material and human—to the peripheries. Jeyifous, however, reverses the dream—future Lagos is a city in which

B Olalekan Jeyifous, Shanty Mega-Structures, 2021

high-rise buildings are derelict structures—rendering the use of verticality deeply ironic: The visual awe of the high-rise is awakened, even as its aspirational tone is negated by the introduction of waste and decay.

Yet while these images deflate the rhetoric of verticality in that regard, they also exploit the imposing vertical axis to augment the presence of the marginalized. The inclusion of shantytowns—just like the poor that they house—draws attention to social fears around increasing urban poverty and can be interpreted as a symbol of resistance against their potential erasure from the future cityscape. According to the WHO Urban Health Report, urban dwellers are set to outnumber their rural counterparts in the near future, with many more people living in informal settlements, meaning that *Shanty Mega-Structures* is less of a reflection of African exceptionalism than we might imagine because the future of global urbanism is the slum.[9] Jeyifous's *Shanty Mega-Structures* thus gestures at how "the future of our planet might be played out on the African continent."[10] In the Anthropocene, Africans are plunged into the center of ecocrises and left to find workable socio-material solutions on their own. Yet by showing the extent to which Western modalities of waste and its management are unfeasible in Africa, it also raises awareness of the fact that our strategies and responses are our own—making the role of Afrofuturism more crucial than ever before.

In comparison to other strategies outlined by the waste hierarchy (like reduce and reuse), recycling lies close to the bottom of the waste hierarchy because it is the most cumbersome way to process waste. It must collect, sort, and reprocess waste into raw materials that can be used in the production of new and more viable products. Ironically, Afrofuturism inverts the hierarchy by exploring a world in which recycling produces the most favorable form of self-expression. Yet, as Till Förster so eloquently explains, for African practitioners, recycling "is not about bringing bits and pieces together into their former, useful lives; it is about alienating them from their former existence and integrating them into another context that lends them another significance and meaning."[11] In contrast to the purposeful assembly of conventional recycling, Afrofuturism approaches recycling as an art of making strange, of de-forming technological waste in unexpected ways such that they give rise to new imaginaries, futures, and identities.

This awareness of waste as a portal to wonder is potently rendered in Fabrice Monteiro's photography, where landfills and dumpsites in Dakar are turned into aesthetic backdrops for majestic and unsettling portraits of individuals. By bringing these subjects into close association with waste, these figures defy our conventional understanding of human beings and bodies; both deformed and reformed through the inclusion of waste, they evoke

C Fabrice Monteiro, The Prophecy, 2025

the wonder at the sub- or superhuman forces that lie beyond our limited human comprehension. Afrofuturism thus reincarnates the "junk status" of othered bodies by bringing together the odd assemblages, surreal selves, and the rapturous awe of the fantastic as a celebration of ontological potentialities for a someone and somewhere beyond the *here* and the *now*.

Similarly, Cyrus Kabiru reassembles e-waste into fantastic and futuristic eyewear. These hyperbolic pieces, however, do not enhance ordinary sight. Instead, they signal the need for an otherworldly vision to confront what lies in front of us.[12] It is often the case that by overriding a default reflex of disgust toward waste, Afrofuturism remains open to a playful dalliance with

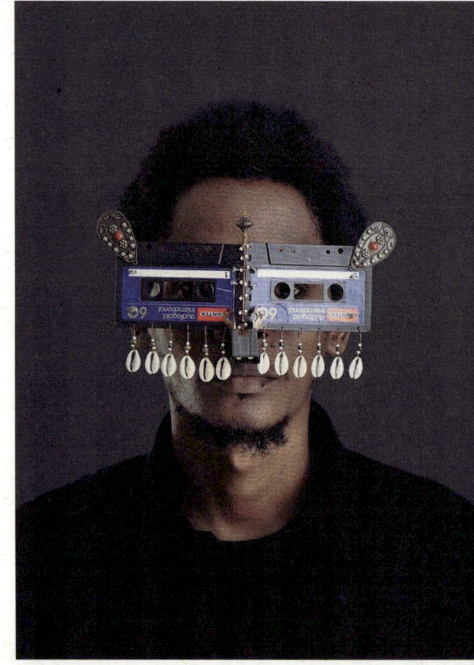

D Cyrus Kabiru, C – Stunners, 2016

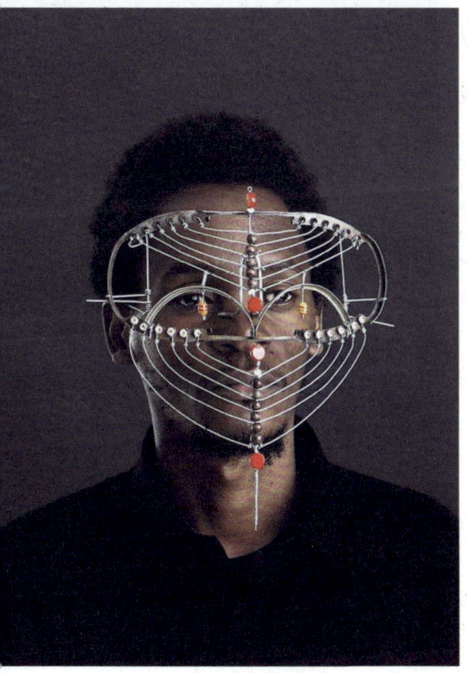

e-waste that allows the machine to break free and become something else entirely.

Here, in this exhibition, Afrane invents his own language to describe this alchemical process; he explores TRONSFORMATION as a method of dissecting e-waste in order to build new and wondrous objects. Tellingly, TRONSFORMATION is about producing the “anti-machine” as objects stray from their original purpose toward a new orientation of purposelessness and aesthetic pleasure. By recycling imaginaries from post-apocalyptic worlds featured in films like *TC 2000* (1993) and *Alita: Battle Angel* (2019), Afrane invites us into a witty appreciation of waste as a repository for potential, not failure.

E Akwasi Bediako Afrane, D²NA – TC24: Visions of the Past, 2024
F Workshop TRONS'R'US by the artist Akwasi Bediako Afrane, 2015

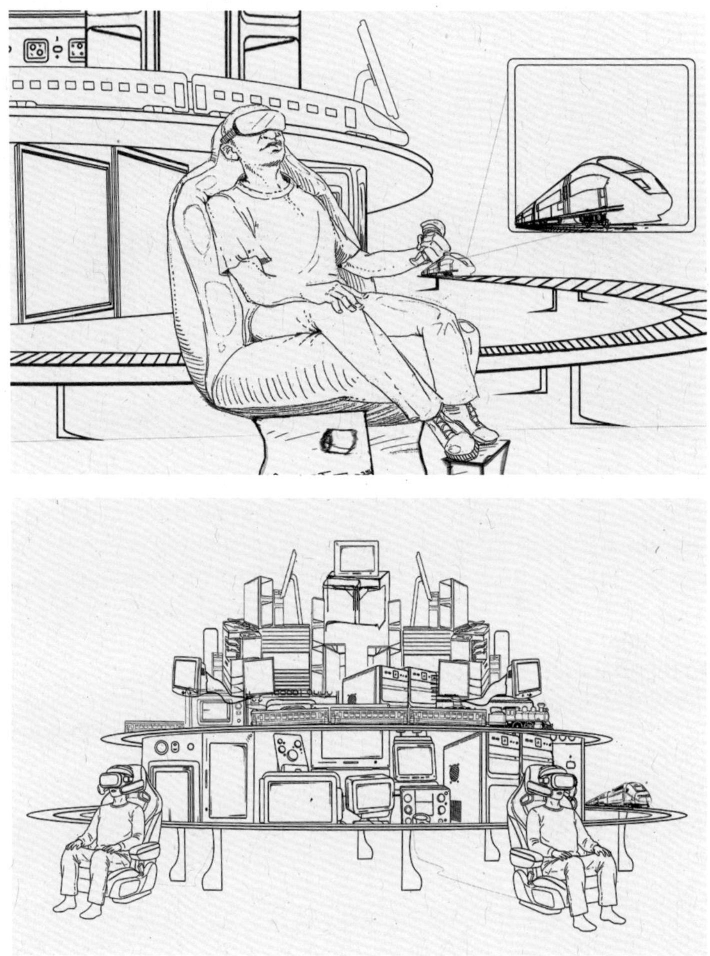

G Akwasi Bediako Afrane, design drawings for TC-2000, 2025

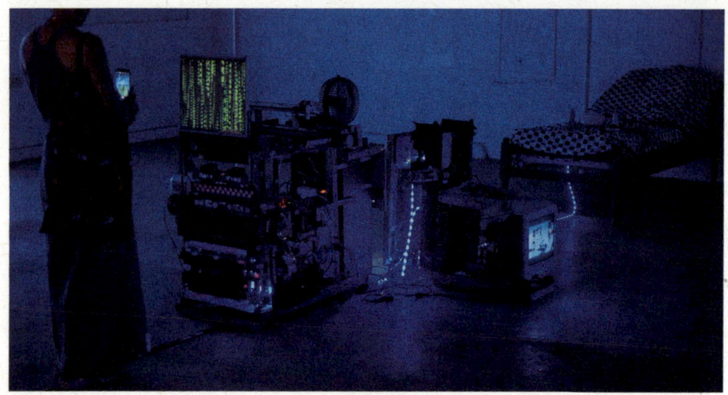

In *TC-2000*, odd connections between found objects emerge, and our notions of technological progress become warped as we witness the past colliding with the present and pointing to future machines we are yet to encounter. One such future machine is also that of the human. Here Afrane envisions a "junk city" where the human is embedded in the machine and becomes a part of its organic evolution and adaptability in the future. For many Africans, the landfill is a new environment of play, as e-waste becomes an imaginative bridge toward African futures to come. By plugging us into the machine, the immersive cityscape of *TC-2000* allows us to experience waste as both weird and wondrous and invites us to ask: Who will we become?

Nedine Moonsamy

Nedine Moonsamy is an associate professor in the English department at the University of Johannesburg. She is currently writing a monograph on contemporary South African fiction and otherwise conducts research in science fiction in Africa. Her debut novel, *The Unfamous Five* (Modjaji Books, 2019) was shortlisted for the HSS Fiction Award (2021), and her poetry was shortlisted for the inaugural National Poetry Prize (2021) awarded by the South African literary magazine *New Contrast*.

Notes

1 Stephanie Newell, *Histories of Dirt: Media and Urban Life in Colonial and Postcolonial Lagos* (Duke University Press, 2020), 120.
2 Newell, *Histories of Dirt*, 121.
3 Henriette Gunkel and Kara Lynch, "Lift Off ... an Introduction," in *We Travel the Space Ways: Black Imagination, Fragments, and Diffractions*, ed. Henriette Gunkel and Kara Lynch (transcript, 2019), 15.
4 This term comes from Cajetan Iheka, *African Ecomedia: Network Forms, Planetary Politics* (Duke University Press, 2021).
5 Newell, *Histories of Dirt*, 11.
6 Baloji, *Zombies*, official short film, Baloji BBL Production, posted April 5, 2019, YouTube, 12:08, https://www.youtube.com/watch?v=pJuZ7qhcKho.
7 Johan Hultman and Hervé Corvellec, "The Waste Hierarchy Model: Disassembling and Reassembling the Socio-Materiality of Waste," *SSRN Electronic Journal* (May 2011).
8 Jack Sullivan, "Trash or Treasure: Global Trade and the Accumulation of E-Waste in Lagos, Nigeria," in "Narratives of the African Landscape: Perspectives on Sustainability," *Africa Today* 61, no. 1 (Fall 2014): 105.
9 "Urban Health," World Health Organization, accessed December 11, 2025, https://www.who.int/health-topics/urban-health#tab=tab_1.
10 Achille Mbembe and Sarah Balakrishnan, "Pan-African Legacies, Afropolitan Futures," *Transition* 120 (2016): 31.
11 Till Förster, "On Creativity in African Urban Life: African Cities as Sites of Creativity and Emancipation," in *Popular Culture in Africa: The Episteme of the Everyday*, ed. Stephanie Newell and Onookome Okome (Routledge, 2014), 38.
12 Iheka, *African Ecomedia*, 134.

Illustrations

A Baloji, Zombies, 2019, film (film stills), © Baloji BBL production
B Olalekan Jeyifous, Shanty Mega-Structures, 2021, digital collage, © Olalekan Jeyifous
C Fabrice Monteiro, The Prophecy, color photographs, 2025, © Fabrice Monteiro
D Cyrus Kabiru, C – Stunners (series), 2016, found objects, © Cyrus Kabiru
E Akwasi Bediako Afrane, D²NA – TC24: Visions of the Past, 2024, Home appliances, electronics, electrical cables, plexiglass, VR headsets, model trains, © Akwasi Bediako Afrane; photo: Tom Little
F Photographs from artist Akwasi Bediako Afrane's workshop TRONS'R'US, 2015, © Akwasi Bediako Afrane
G Akwasi Bediako Afrane, draft drawings for TC-2000, 2025, pencil on paper, © Akwasi Bediako Afrane
H Akwasi Bediako Afrane, ALTERED KARBONS, 2025, © Akwasi Bediako Afrane; photo: Anwar Sadat Mohammed

BITTERFELD, OR THE LAWS OF WASTE

Oliver Schlaudt

"B. is the dirtiest city in Europe." This sentence was once intended as the beginning of a journalistic report on Bitterfeld, a report whose existence was an impossibility. It was only contrived in a novel, and even in the story told there, it had to remain unprinted. Today, the small town in central Germany is still rated as one of the world's most contaminated areas, the earth beneath it saturated with 200 million cubic meters of highly polluted groundwater. The technical term for this is a "contaminated mega-site." The composition of the toxic sludge is a matter for chemical analysis. But what of its cultural significance? What does the polluted landscape have to say to us? What does it say about us? What takeaway truth does it bring home to us? This is not something we can find out in the laboratory, but rather through works of art—such as Monika Maron's novel *Flugasche* (quoted above) or Anna Zett's video work *Freiheit 3*, named after one of the Bitterfeld dumps.[1] This may be the dimension that Zett explores when she writes messages to nature, sprayed onto the slippery slope of a mountainous gravel heap—cut together, in the video montage, with scenes of East German citizens grappling with the reality of their obsolete country—before immediately obliterating them again.

So let's face up to the challenge and ask ourselves some basic questions: What is waste? What is pollution? Where do they come from, and where are they leading us?

A Anna Zett, Freiheit 3, 2019

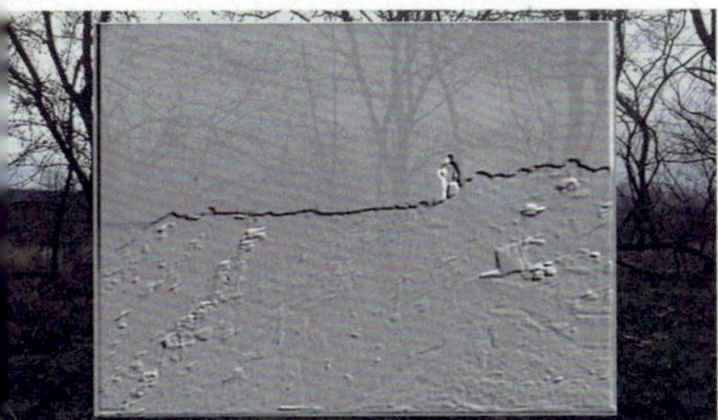

Let's examine the literature. "Pollution is natural," we read—a sentence that is bound to bewilder. Was it written by a waste denier? In fact, it was penned by the American biologist Lynn Margulis in one of the works in which she popularized Gaia theory.[2] According to this theory, the earth has the attributes of a living organism: It is a gargantuan living being, more than three billion years old, following its lonesome trajectory through space. Just as we are made up of cells, Gaia consists of countless living beings that lay down a gossamer-thin wafer of biosphere around the planet's mineral core. And for this cosmic being, waste is most certainly a problem. One of its constituent elements, humanity, has begun to spew out waste in some places—waste that poisons Gaia from within. One might almost say that Gaia is suffering from humanity as one would suffer from a disease.

LIEBE UMWELT
ES GIBT DA ETWAS
DAS ÜBRIG
GEBLIEBEN
IST

LIEBE
ELT
NICH

"Pollution is natural"—How does that fit the picture? Humanity's process of industrialization is, after all, not the first instance of poisoning in Gaia's medical records. Margulis cites an event from Gaia's youth, about 2.4 billion years ago, which has gone down in the annals of natural history as the "great oxygen catastrophe." Back then, cyanobacteria further evolved their ability to harness sunlight as an energy source using photosynthesis. Prior to that, these primitive bacterial organisms had relied on obtaining energy from the young planet's minerals and gases, which condemned them to a frugal existence at the bottom of the oceans. The vast quantities of sunlight that flooded the earth remained unused. Up until the invention of photosynthesis, which catapulted life to unprecedented energetic heights. But there was a snag, even if it later turned out to be a success story: As they feasted on sunlight, the cyanobacteria excreted large amounts of oxygen as a by-product of photosynthesis. This brings us already to the first law of waste:

LAW 1, OR THE LAW OF LIFE:
ORGANISMS GENERATE WASTE.

Life demands a constant flow of energy and matter, which brings into being a dialectic of self and other. In order to be and remain themselves, organisms must absorb, assimilate, and excrete the other, that which is extrinsic to themselves.

However, the oxygen that the cyanobacteria began to excrete was highly toxic for most primitive organisms on Early Earth. For a while, the planet was able to absorb the lethal substance. For example, it was soaked up by the iron on the earth's surface through the process of rusting, as is evident in the rust-red layers of the banded iron formations that occur in many places on earth. But, at some point, these sinks became saturated, and there was a rapid buildup of oxygen in the atmosphere, causing young life on earth to die off on an unprecedented scale. As oxygen broke down the methane in the atmosphere, the greenhouse effect collapsed, taking with it the warmth it generated. Starting from the earth's poles, glaciers grew and probably engulfed the entire globe, which fell into a hibernation that lasted four hundred million years, a phenomenon known as "Snowball Earth."

We know that this was not the end of the story—our own existence is a testament to life's continuing evolutionary processes. And life did not just find a way to immunize itself against the toxic effects of oxygen, it actually discovered in it the basis for a completely new and highly efficient means of harvesting energy—oxygen respiration, in which atmospheric oxygen is used to tap the chemical energy contained in the physical substance of other organisms: i.e., plant and animal matter. As Margulis encapsulates it, "New wastes test life's tolerance and stimulate life's creativity... The cyanobacteria's waste became our fresh air."[3]

So, pollution is natural after all. The seeming absurdity of this is dispelled the moment you remember that in nature, as Margulis boils it down, "one organism's waste is another's food."[4] The air we breathe—for us, a symbol of life and purity—is excreted by plants as a waste product. We breathe fecal matter!

Thus, nature creates its own recycling mechanisms. We should beware, of course, of putting an overly harmonious spin on this, because "nature," in truth, is never quiescent. It is in a constant evolutionary ferment, creating something new, including new waste that first has to be made usable. After eons of cold, the further processing of oxygen opened the way for higher life forms. Where before only bacteria had any prospects, complex, multicellular life forms were now able to establish themselves—including, at some point, we humans, thus beginning a new chapter in the history of waste.

Humans are classed as "cultural animals" equipped with a treasure trove of learned behaviors that are not governed by instinct and can be passed on from person to person and from generation to generation. This affords us greater flexibility in adapting to external circumstances and thus makes it possible for us to colonize new habitats. Since we not only learn behaviors but can also modify, refine, and co-opt them—that is, build on the prior knowledge of our elders—there have been staggering developments in human culture through history, from "the hand ax to the nuclear power plant," which are as instrumental in our lives as our basic biological hardware: cultural evolution.

There is one thing missing, though, from the success stories emerging from biology and archaeology: As our culture becomes more complex, so too does our waste. "From the hand ax to the nuclear power plant" also means from a few remnants of stone and bone to radioactive waste.

LAW 2, OR THE LAW OF PROGRESS: WASTE HAS ITS OWN BIOLOGICAL AND CULTURAL EVOLUTION.

This brings us closer to the nub of the problem. Grasping the nettle is not just a matter of getting our heads around waste; we must also give thought to how we deal with it. Zooming out to view the whole of modern human culture alerts us to a strange asynchrony: What we produce is becoming ever more complex, and so the waste we produce is becoming increasingly complex to match, yet we still drop this waste behind us as we go as if it were excrement or a few shards of pottery. We are ultramodern in terms of the waste we produce but millennia behind the curve in how we dispose of it. Since snowballing cultural development, fueled by its own momentum, has outrun biological evolution, the biosphere that girdles us has no chance of contriving methods of composting or recycling.

The chemical industry came to embody this metabolic logic, manifesting as a strange kind of creature whose traces can still be found in Bitterfeld. It gobbled up coal, which it dug out of the earth, its claws armed with bucket-wheel excavators, and dropped its feces into the worked-out pits of the strip mines, liter by liter, kilogram by kilogram, with highly toxic chlorinated hydrocarbons leading the way, including not only chlorobenzenes, chlorophenols, and hexachlorocyclohexanes—which belong to the most problematic of any group of substances—but also nitro compounds, azo dyes, organic filter cakes, phosphorus pentasulfide, heavy metals aplenty, and much more.

Some five thousand chemicals are thought to have been used and produced in Bitterfeld. They have now all found their way into the city's subsoil, where many more may also exist, precipitated by the cascades of chemical reactions that have potentially been triggered down there. The danger inherent in such a situation became acute when the coal stopped being mined. The pumping required to keep the pits dry had lowered the groundwater by several meters across a large area. Now, however, the bottoms of the disposal sites are standing in water, and the toxins are gradually leaching out. In some places, they are infiltrating the basements of buildings—the elementary school in Greppin, for example, which had to be closed after it was found to be contaminated with the

carcinogen vinyl chloride. The Bitterfeld plume is inching its way toward the Mulde River, which has already acted as a conduit for some of the waste from the area, disgorging it into the Elbe and eventually into the North Sea.
We encounter here two further laws relating to waste:

LAW 3, OR THE LAW OF DISSIPATION:
WASTE DISPERSES.

Once waste has been released into the environment, it tends to diffuse and becomes increasingly difficult to capture. As we know from pesticides, heavy metals, and microplastics, waste can now be found everywhere, even in the wildest, most inaccessible areas of the earth. Anyone hoping that the toxins would at least keep being diluted in the process and so become harmless has failed to take account of an opposing tendency in the biosphere:

LAW 4, OR THE LAW OF ACCUMULATION:
THE PROCESS OF LIFE CAUSES WASTE TO BECOME CONCENTRATED AGAIN
AND THEREBY RETURNS IT TO US.

General dissipation is counteracted by the process of life itself. Organisms take in waste products with their food and accumulate them in their tissue, so that they find their way back into our food chain in concentrated form.

C Anna Zett, Untitled, 2024

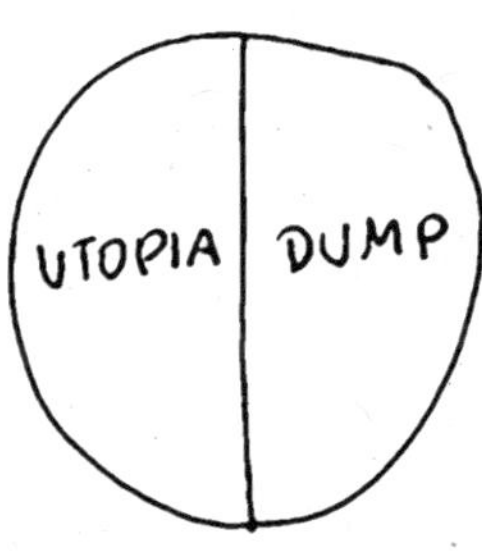

Bitterfeld toxins can still be detected, even today, in fish from the German Bight.

This makes Bitterfeld a pressing problem—a problem that should not have existed. In a text accompanying her video works, artist Anna Zett elaborates on the mindset underpinning the establishment of landfills—these dumps were treated as a utopian dream, proverbial non-places where the laws of nature and the laws of waste were suspended: "The landfill is expected to operate according to rules that fundamentally differ from the rules of ecology. It is based on the phantasm of isolation, on the utopian idea that a non-place can be sustained."[5]

Every dump is a utopia, because isolation cannot function in a world built on connections. Zett also reflects this in a contemporary historical context: Every utopia needs a dump where it believes it can get rid of what is bad, dispose of the other, of what is extrinsic to itself. These ideas, which take the impossibility of utopia seriously, are expressive of a sense of disillusionment.

And what are we to do with the inherited waste that simply cannot be disposed of?

After the fall of the Berlin Wall, when an entire country was disposed of on the great ash heap of history, there was a belief that everything could be done better, and the landscape could be made to flourish again; millions of euros were invested, and modern technology deployed. Capitalism makes things clean. “Bioremediation” is a way of giving the biosphere a leg up, as it were, and bacteria are artificially created that can feed off the toxins. Optimized microorganisms can break down petroleum, diesel, aromatic hydrocarbons, and even some pesticides and are capable of immobilizing heavy metals. But how can you cope with a cocktail consisting of five thousand substances? Here’s what the engineers have to say on their return from a fruitless attempt to carry out the big clean-up: “While the urge to control and clean up mega-sites is understandable, a more realistic response to these massive and persistent sources of contamination is to acknowledge their position as new features of the socio-ecological landscapes in which they are located.”[6]

The dump was part of a utopian dream, as was its technological remediation. We need to accept contaminated sites as part of the landscape and learn to live with them. Now that disposal has failed, a new, antithetical disposition must take its place: From seeking to “dis-care” (a literal rendering of the German term for disposal, *ent-sorgen*),

D Anna Zett, Endarchiv, 2019

DIE DEPONIE
VERGISST NICHTS

we must learn to take care. And so Bitterfeld remains the utopian non-place that the chemical industry had intended it to be—it is just that the auspices under which it was conceived have been flipped around and impossibility has become inevitability. We must live with the contaminated sites. Bitterfeld is thus becoming a place of the future—where we can practice the "arts of living on a damaged planet."[7]

Oliver Schlaudt initially trained as a physicist and now works as a philosopher focusing on issues relating to technology, politics, ecology, and economics. In 2024, his book *Zugemüllt: Eine müllphilosophische Deutschlandreise* was published by C. H. Beck. This was followed in 2025 by the UTB publication *Wirtschaft im Anthropozän: Grundbegriffe der ökologischen Ökonomie*.

Notes

1 Monika Maron, *Flugasche* (Fischer, 1981).
2 Lynn Margulis, *Symbiotic Planet: A New Look at Evolution* (Basic Books, 1998), 121.
3 Margulis, *Symbiotic Planet*, 121.
4 Margulis, *Symbiotic Planet*, 119.
5 Anna Zett, "Dispose, Discard, Discare: Environmental, Political and Emotional Wastelands in the German Democratic Republic," *ArtMargins Online*, September 3, 2025, https://artmargins.com/dispose-discard-discare/.
6 Mario Schirmer et al., "A Socio-Ecological Adaptive Approach to Contaminated Mega-Site Management: From 'Control and Correct' to 'Coping with Change,'" *Journal of Contaminant Hydrology* 127, nos. 1–4 (January 2012), 103.
7 Anna Lowenhaupt Tsing et al., eds., *Arts of Living on a Damaged Planet: Ghosts and Monsters of the Anthropocene* (University of Minnesota Press, 2017).

Illustrations

A Anna Zett, Freiheit 3, video (film still), © VG Bild-Kunst
B, D Anna Zett, Endarchiv, 2019, © VG Bild-Kunst
C Anna Zett, Untitled, drawing on paper, 2024, © VG Bild-Kunst

E5104L
011310
5BC
-NUGQ
UBUANK1

UNGOVERNABLE WASTE:

FROM COLONIAL TEMPORALITIES OF VIOLENCE TO REPAIR

Evelyn Wan

Every technological gadget is a ticking clock, with an inevitable countdown to its obsolescence. Sometimes these gadgets die because of hardware issues. Sometimes they are outpaced by software development and are rendered useless. Where have your old gadgets ended up? My old Nokia E65, a purple slide phone with only 50 MB of built-in memory, still lies untouched in my drawer as a keepsake. With fond memories of my youth attached, I am not ready to toss my Nokia in the bin, even though it has long been replaced by fancier smartphones.

Numerous electronic devices are trashed every day and trafficked to e-waste dumps in different parts of the world. The countries of West Africa, specifically Ghana and Nigeria, receive the highest percentage of European e-waste and have the most dangerous informal processing conditions. Southeast Asia (Malaysia, Vietnam, Thailand, Philippines, Indonesia) became an important player after China officially banned waste imports in 2017. Meanwhile, India, Pakistan, and Bangladesh have a significant informal e-recycling industry. Latin America also recognizes a growing problem in receiving e-waste shipments, as the developed world pushes their trash elsewhere. E-waste is not only sent officially to different processing sites but is also funneled into black markets, where it vanishes into smuggling networks and into ports and places where regulation is lax.

Waste networks demonstrate the persistence of colonial logic in the racialized disregard for waste laborers elsewhere, and there is often criticism of how developed countries use them to export or expel their waste to other parts of the planet. In his 2004 book *Wasted Lives*, Zygmunt Bauman argues that waste is a defining feature of modern life and progress. For modernity to keep moving forward, certain people must take on the role of collecting and recycling society's trash. According to Bauman, progress is driven by our constant craving for the newest things, which means that we are always throwing away what has become old or unwanted.[1] Will we ever step away from this vicious cycle of production, consumption, and disposal, a capitalist rhythm resonating with colonialist, exploitative overtones? What would help us switch to circular sustainable economies, characterized by slower rhythms of regeneration and repair?

To address the subject of the planet's e-waste hotspots, my essay focuses on Hong Kong, a postcolonial city where citizens generate more than double the global average of e-waste (20.2 kg vs. 7.3 kg per capita per year), while receiving an additional 1,000 metric tons of e-waste daily from other developed regions for further processing.[2] Situated next door to China, which remains the world's largest e-waste processor, the city embodies two temporalities of violence overlaid on each other—one of vicious consumption, where young hip urbanites would not be seen dead with an outmoded phone, and the other of cruel dispossession, where the poor and undocumented end up processing e-waste in unsafe conditions in border zones. The following anecdote probes the implications of post-colonial e-waste politics and the ungovernable nature of the problem.[3]

E-WASTE, IN MY OWN BACKYARD

In 2017, as part of an academic conference on digital infrastructures in Hong Kong, I joined my colleagues on a field trip to illegal e-waste processing sites in Hong Kong. Up in the far north, in the border zone between mainland China and Hong Kong, landlords looking for a quick buck were facilitating the temporary processing of smuggled e-waste.

China's imposition of ever-stricter regulations in the 2010s, followed by a ban in 2017, meant that Hong Kong, Southeast Asia, and South Asia grew as hotspots for e-waste. Hong Kong also functioned as a port for illegal exports to the mainland, while turning into a place where e-waste is processed behind closed doors. This illegal economy relied on undocumented workers who operate outside legal protections.

We got out of the van after a long trip to a village in the New Territories with a name I can barely pronounce. The moment I alighted, I saw severed wires, chipped circuit

A Illegal recycling site in the border area of the New Territories, Hong Kong, 2017

Workers in an e-waste recycling facility, seen through a peephole, 2017

boards, and broken pieces of plastic hardware poking out of the ground as if they were part of the local flora. Had we entered a post-apocalyptic world where the soil was simply mixed with inorganic matter? There were high sheet-metal walls everywhere, fencing off the e-waste processing sites. A threatening graffitied message read, “Private Land: No Parking,” and a series of CCTV cameras, installed for surveillance purposes, rose up above the metallic walls. On one wall, a half-torn poster promoted an app for the global waste trade, a platform where businesses could bid on waste shipments and participate in the recycling trade. I noted the Hong Kong business address, but the listed phone number used a mainland Chinese area code.

I peeked from between metallic barriers and bent down to look through a hole. Inside were rows and rows of bundled machines, from printers to DVD players, waiting to be disassembled. A middle-aged woman, likely a migrant worker from mainland China, stood in a gentle slump. Her face was covered by a mask, and she wore a large wide-brimmed hat. She was on her own, working close to us, half-hidden from view by the stacks of CPUs surrounding her workstation. I heard the sound of mechanic tools in the distance—electric drills, perhaps, and the steady beat of a hammer, as if someone were trying to break open a metallic object.

How was it possible for this border area to be completely zoned off and unregulated? How did it reflect the colonial

E-waste in front of a recycling facility, 2017

violence suffered by the dispossessed bodies who worked in these sites? A short excursion into the postcolonial politics of this area is necessary.

Hong Kong's New Territories enjoys an exceptional status in terms of land rights and governance, a colonial legacy from British occupation. When Britain acquired the New Territories in 1898, it inherited a rural society with its own set of customs, clans, and land ownership systems. The British decided to recognize certain customary practices and granted special rights to indigenous villagers to appease them. Rural power structures were instituted, so that the indigenes had sufficient political representation.

The colonial government relied heavily on rural intermediaries to negotiate land development projects, which eventually resulted in a powerful rural elite that exerted considerable influence in directing affairs in the New Territories. Indigenous male villagers have also had the right to each build a small house in the vicinity of their village since the 1970s. Over time, villagers have gathered to monetize this right. Village land would be put together, and deals negotiated with developers to unlock profits. Even after Hong Kong was handed back to China, rural elites such as village leaders maintained the power of governance, and government officials need to gain their favor in order to put forward new policies in the New Territories. There has historically been weak government oversight in these areas, which has also enabled the emergence of local triads and other informal power structures that facilitate unregulated and potentially illegal activities on private land. This phenomenon has been described by local activists as a collusion between four parties: government, businesses and developers, rural elites and landlords, and the triads.[4]

Even though this scenario is very specific to Hong Kong, one can observe how poor governance is a result of the colonial legacy. The combination of capitalist desire and rural power grabbing has turned village plots in the New Territories into no-man's-land. The lack of formal regulatory frameworks opened up possibilities for illegal e-waste processing through the exploitation of loopholes. E-waste enters the scene as an ungovernable object and meets unscrupulous landlords and businessmen looking for cash.

Yet those who bear the brunt of the toxic processing are the poor and the dispossessed. Through local contacts in the New Territories, we met refugees from South Asian countries who were stuck in limbo, forbidden to work but unable to subsist on government subsidies. In desperation, they turned to these black-market jobs. They were reluctant to say anything about their experience and were scared that we would inform on them, as this could result in criminal prosecution, fines, jail time, and deportation.

The refugees left before I could ask if they understood the slow violence of e-waste processing, if they were aware of the toxic dust and carcinogens that came with cracking open devices and burning parts for valuable metal extraction. I wanted to warn them that studies have linked exposure to such pollutants with cancer, DNA damage, lung disease, and increased rates of birth defects and miscarriages. Merely wearing face masks and gloves could not protect them from harm.

The field trip left a bitter taste in my mouth, and I was confronted with the realization that toxins were leaching into my hometown from these illegal sites, knowing from the literature that e-waste processing contaminates groundwater, and toxins end up in the food chain.[5] This was most vividly captured by research on China's cancer village, Guiyu, notorious for its streets of family-run e-waste processing workshops. Toxins have remained in the soil, affecting crops even 400 km away, which tested positive for cadmium contamination years after the e-waste operations had closed down. A 2019 Chinese report recommended soil repair in former e-waste areas, as animal and plant life is still suffering from e-waste toxins almost twenty years later.[6] The aftereffects of e-waste persist through the ecological chain, impacting plants, animals, soil, and humans, a phenomenon Jisha Menon terms "toxic colonialism."[7] I also see this as a colonial imprint of violence, one that unfolds long after the colonizers have left, but whose effects persist, from structures of poor governance to ecological damage to blatant disregard for certain racialized lives.[8]

D E-waste in front of a recycling facility, 2017

Since January 2025, Hong Kong has strengthened import and export control on electrical and electronic waste, and new schemes have been implemented to facilitate the recycling of products. More companies have been set up to handle the large volumes of e-waste, in a legal and regulated manner. I have not had a chance to revisit the border zones since the new rules came in to see whether these sites have closed down. Yet history tells us that when a place introduces and reinforces stringent rules, excess e-waste is simply pushed to other hotspots in neighboring countries, just as when China's 2017 import ban resulted in cross-border smuggling in Hong Kong. There is always another well-connected port to take on smuggled shipments, from Haiphong in Vietnam to Lagos in Nigeria. The Hong Kong example is also proof that even advanced economies are not immune to e-waste toxicities. The only way for us to move beyond ungovernable e-waste is to break away from the toxic cycle of "progress"-driven consumption and the habits of waste that ensue from it. We must resist the ticking clocks of obsolescence and turn toward sustainable cyclical visions of time.

TOWARD REGENERATION AND REPAIR

In a perfectly cyclical system, there is no waste. In nature, all waste materials are reabsorbed by the ecosystem for regeneration. Death becomes compost, and compost becomes fodder for life. It is only in our severed relations with nature and with the earth that we continue the colonialist attitude of extraction and expulsion and deny the effects of toxicity as long as they happen elsewhere.

I am inspired by the "broken world thinking" espoused by media scholar Steven Jackson: "What happens when we take erosion, breakdown, and decay, rather than novelty, growth, and progress, as our starting points in thinking through the nature, use and effects of information technology and new media?"[9] Taking breakdown seriously, how might we activate "repair," not only to reduce e-waste by fixing broken gadgets but also to take on the larger normative and ontological project of repairing and redressing the colonial durations of violence that persist in our world?

Breakdown may be inevitable, but e-waste is not. We need to face the violence embedded in our consumption and disposal patterns and cultivate new habits. We ought to see waste processing and recycling not as work for the dispossessed in developing countries but as valued labor essential to our supply chains. We must recognize these laborers as the repair workers of our digital infrastructures. Rather than living wasted lives as victims of the excesses of progress, they remedy the lack of recyclability. Their work essentially breaks down waste materials, countering disposable designs. On the frontlines, they mend the unmendable—the gulf between rampant techno-capitalism and the forever-deferred sustainable future.

E-waste is perhaps only ungovernable because of our ungovernable desires. For it is these desires that generate impossible volumes of waste. To Jackson, foregrounding breakdown and repair once again embeds moral relations in the world of technology. In our current times, renewed dedication to an ethics of mutual care and responsibility is urgently needed. Colonial violence may haunt our interconnected world, but we can commit to the slower rhythms of repair, where we fix and reimagine the world, one act of care at a time.

Evelyn Wan

Evelyn Wan is an artist-scholar and dramaturge. She is assistant professor in media, arts, and society in the Department of Media and Culture Studies at Utrecht University, where she coordinates the Master of Arts and Society program and works on interdisciplinary curriculum innovation in the domain of Humane AI. She is a founding member of the Hong Kong–based artistic collective If Time's Limited.

Notes

1 Zygmunt Bauman, *Wasted Lives: Modernity and Its Outcasts* (Polity, 2004).
2 Gengze Liao et al., "Assessing Neurobehavioral Alterations Among E-Waste Recycling Workers in Hong Kong," *Safety and Health at Work* 15, no. 1 (2024).
3 I borrow the notion of ungovernability from Rolien Hoyng, "Logistics of the Accident: E-Waste Management in Hong Kong," in *Logistical Asia: The Labour of Making a World Region*, ed. Brett Neilson et al. (Palgrave Macmillan, 2018).
4 This is known colloquially as government-business-landlord-triad collusion.
5 Ironically, round the corner from the e-waste sites in Hong Kong were eco-farms, popular with families looking for a fun day out in the countryside, usually to pick organic strawberries. I could not help but wonder whether organic vegetables with hefty price tags in fact grew from soil with heavy-metal contamination.
6 Yunjiang Yu et al., "Health Implication of Heavy Metals Exposure via Multiple Pathways for Residents Living near a Former E-Waste Recycling Area in China: A Comparative Study," *Ecotoxicology and Environmental Safety* 169 (March 2019).
7 Jisha Menon, "Toxic Colonialism and the Gesture of Generosity," *Performance Research* 23, no. 6 (2018).
8 See also Evelyn Wan, "Laboring in Electronic and Digital Waste Infrastructures: Colonial Temporalities of Violence in Asia," *International Journal of Communication* 15 (2021).
9 Steven J. Jackson, "Rethinking Repair," in *Media Technologies: Essays on Communication, Materiality, and Society*, ed. Tarleton Gillespie et al. (MIT Press, 2014), 221.

Illustrations

A–D All illustrations in this essay © Evelyn Wan

02641F5 LI0122B1B4V3. ON

ART, SPECTACLE, AND THE STAGE OF EXTRACTION:

GREENPEACE AND THE AESTHETICS OF RESISTANCE

Annabel Keenan

From its inception in the early 1970s in Vancouver, Canada, Greenpeace understood that the case for protecting the planet could never be waged solely in courtrooms, laboratories, or diplomatic chambers. The environmental movement has always been acutely visual, and art is a powerful tool with which to give tangible form to complex and difficult issues. In the early years of Greenpeace, just as today, examples abound of the work of activism being bolstered by creative documentation and artistic interventions. At times, these visuals can be crucial to the cause at hand. Take, for example, sea-bound activism, which often occurs offshore or in remote regions. The photographs of activists in flotillas dwarfed by whaling ships shed light on important advocacy and disseminate the dangerous work being done to protect the ocean and its inhabitants. Art can force a reckoning.

What followed Greenpeace's early environmentalist appeals was almost inevitable: a deepening collaboration with artists whose work could insert ecological urgency into public consciousness and become a catalyst for action.

Over five decades, Greenpeace has partnered with artists who move fluidly between the spheres of the gallery and the street, the museum and the media cycle. The organization leverages art's capacity for rupture and its ability to tap into the public's imagination. The nonprofit has collaborated with well-known artists like Judy Chicago, Anish Kapoor, Nicolás García Uriburu, and Banksy, revealing how aesthetics can function as strategic insurgency. Capitalizing on the public sphere, these artists bring activism straight into the viewer's world, forcing them to engage with issues they might not understand or want to consider.

THE BIRTH OF ENVIRONMENTAL ART

As an artistic genre, the seeds of the environmental art movement were planted in the years that followed World War II. Engaging with climate concerns and other areas of contention, art became a tool for activism as artists gave visual expression to the grievances of the day, which often centered on war, politics, and social issues like education. Public concern for the environment gained traction in the 1970s, with the first Earth Day celebrated in 1970 in the United States, and the United Nations Environment Programme launching its annual World Environment Day in 1973.[1] Phrases like "only one Earth" entered popular culture, and famous environmentalists like David Attenborough and Jane Goodall became not just beloved figures but trusted voices too.

Environmental art as a theoretical concern emerged in this landscape, pioneered by people like Agnes Denes, whose wheat fields planted in locations emblematic of capitalism, such as locations close to Wall Street, New York City, decried greed and inequality. At the same time,

A Nicolás García Uriburu, Basta de Contaminar, 1999

art left the walls of the white cube and took to the streets and the public realm through land art, as epitomized by Robert Smithson's *Spiral Jetty* (1970) rock installation on the shores of the Great Salt Lake in Utah, and in the form of graffiti across urban spaces. Argentinian artist Nicolás García Uriburu embraced the union of public art, environmental activism, and spectacle, most famously with his iconic 1968 Venice action in which he dyed the Canal Grande a phosphorescent green with a nontoxic powder, insisting pollution be seen.

Uriburu knew that visibility could provoke accountability. He understood then that ecology can be political theater: the city becomes a stage, the water itself a protagonist. A synthesis of beauty and alarm, the urgency of his method remains a blueprint for activist art today.

What made Uriburu's practice radical was not merely his gesture but his insistence on continuity. Underscoring the enduring problems of waste and pollution globally, he repeated similar dye interventions across continents, in the Seine in Paris, the East River in New York, the Rhine in Düsseldorf, and the Plata in Buenos Aires, linking sites of privilege and pollution. His actions prefigured the language of globalization that would later define environmental discourse, asserting that contamination, like capital, crosses borders with ease.

Uriburu was a natural ally for Greenpeace. His work was less about the representation of nature and more

B Nicolás García Uriburu, Coloración, Canal Grande, Venice, 1968

about confrontation and exposing the complicity of cities and governments in environmental neglect. His collaborations with Greenpeace in 1999 and 2010 in Latin America extended that logic outward, combining water interventions with activism around deforestation and river rights. For these, he hung a banner bearing the slogan *Basta de contaminar*, meaning “Stop polluting” in Spanish, over the Riachuelo River in Buenos Aires, drawing attention to the toxicity of the river, one of the most polluted waterways in Argentina.

The 2010 iteration, “Riachuelo: 200 años de Contaminación (Riachuelo: 200 years of pollution)” a nod to Argentina’s bicentennial, was both an elegy and a warning. Uriburu re-envisioned his Venice, dyeing the river green, a ghostly echo of his earlier performance framed as an indictment of political inertia. The continual pollution of the river underscored the disappointing failure of those in power to take action. Even if we can see pollution, even if the rivers are turning green, the necessary steps to clean the waterways have not been taken, either then or now.

For Greenpeace, Uriburu’s work offered a model for the power of aesthetic repetition, a reminder that environmental crises do not unfold as singular events but as sustained conditions of neglect and waste. His green rivers became a recurring monument, a visual refrain that stitched together decades of activism. Where traditional monuments commemorate victories, Uriburu's commemorated failure. In doing so, he inverted the monument's function: instead of celebrating the past, his works demanded a better future.

Uriburu's ecological aesthetics also anticipated the globalized environmental campaigns that Greenpeace would later perfect: performances staged for maximum photographic impact, interventions calibrated for the news cycle, and gestures that collapse protest, ritual, and art into a single communicative form. Long before hashtags, livestreams, and viral videos, Uriburu understood that the environment's defense was also a struggle for visibility. His legacy is an indelible feature of Greenpeace's visual strategy: to make pollution not only legible but unforgettable.

THE POSTER AS PROPAGANDA AND THE DISTRIBUTED MONUMENT

Not all Greenpeace collaborations with artists are as grandiose as Uriburu's. Before the age of social media, when images could not yet be easily disseminated, the nonprofit benefited from the physical proliferation of images in the streets, partnering with artists on reproducible and sharable artwork. In 1980, Judy Chicago's print *Rainbow Warrior: A Tribute to the Efforts of Greenpeace* reframed the organization's eponymous ship as a mythic protagonist in a struggle against ecocide. Chicago's work was unabashedly agitprop: a poster meant not for white walls but for telephone poles and student union pinboards.

Likewise, in the early 2000s, infamous artist Banksy created *Save or Delete* for a Greenpeace anti-deforestation campaign, co-opting familiar children's characters and setting them within clear-cut wastelands, thus collapsing nostalgia into disillusionment. The controversial legal frictions were part of the piece: Environmental damage is often licensed by bureaucracy, and so the artwork's circulation was itself contested terrain.

Another significant collaboration emerged in 2015, when American street artist Shepard Fairey partnered with Greenpeace on its Save the Arctic campaign. Fairey, whose graphic visual language is synonymous with bold political messaging, produced a series of posters and digital works urging a global halt to Arctic oil drilling. His central image of a polar bear perched on a shrinking iceberg, encircled by industrial silhouettes functioned as both warning and rallying cry. Distributed by Greenpeace internationally, the artwork appeared as wheatpastes, prints, protest materials, and widely circulated digital graphics, allowing the image to migrate across public space and social media.

In 2021, Greenpeace marked its fiftieth anniversary by launching a mural project, commissioning artists includ-ing the Swiss duo Queen Kong to transform urban facades into climate manifestos. These works function somewhere between celebration and alarm. They are murals as social sculpture grounded in local environmental stories yet tethered to a planetary narrative.

The tactic is clear: Replace the verticality of museum walls with the democratized visibility of the street. The city itself becomes an exhibition platform—the audience, involuntary but implicated.

In today's world, social media has become a powerful tool for reaching global audiences, and Greenpeace has wisely capitalized on the potential of a viral image and digital campaign. In 2020, Greenpeace collaborated again with Judy Chicago on the #CreateArtForEarth campaign, along with American street artist Swoon (Caledonia Curry) and actor and activist Jane Fonda. Here, the artists abandoned singular authorship in favor of participatory ecology—thousands of artists submitting works that circulated through digital commons and public space—underscoring the power of collaboration and collective action. This collaborative ethos and common goal signaled a broader evolution: from the aura of the singular artwork to the distributed monument, collectively authored and infinitely reproducible.

Swoon's large-scale wheatpastes for the #CreateArt-ForEarth initiative were particularly successful in spreading the organization's message far and wide. Easily printed, installed, and shared, the project capitalized on the proliferation of images in everyday life. With #CreateArt-ForEarth, the street and the internet became not only sites of resistance but also conduits for collective imagination.

INTERVENTION AT THE SITE OF HARM

The case for environmental action is one that is hard fought, facing climate change deniers and powerhouse corporations with resources to change narratives and silence opposition. In 2020, British artist Fiona Banner (also known as The Vanity Press) installed a 1.25-ton sculpture called *Full Stop* on the doorstep of the UK's Department for Environment, Food & Rural Affairs (Defra). The work was in support of Greenpeace's activism to stop bottom trawling, a form of commercial fishing in which giant, weighted nets are dragged along the seafloor. Catching all kinds of creatures and damaging habitats, the barbaric practice results in significant loss of ocean life.

Banner's sculpture blocked the entrance to the office, forcing Defra to confront the message and its clear directive. Bringing the work directly to its front door, the work re-framed the government not as a site of scientific neutrality but as a protagonist in a looming ecological crisis.

No industry is a bigger target than that of fossil fuels. In 2025, Anish Kapoor and Greenpeace attacked oil and gas with an artwork called *Butchered*. In an image that verged on the visceral, Kapoor attached a monumental canvas

Artist Fiona Banner aka The Vanity Press with Klang Full Stop outside DEFRA in London, 2020

D Anish Kapoor, Butchered, 2025

stained in a lurid, blood-like hue to the metallic flank of a Shell gas platform in the North Sea. Adorning the very infrastructure of extraction, the canvas made the connection between industry and brutality unmissable. Installed by Greenpeace climbers in an act that bordered on infiltration, the work transformed the platform into a colossal billboard for climate violence.

Like Banner's work, Kapoor's intervention didn't just critique power, it literally clung to it. The artist and non-profit's rejection of symbolic gestures in favor of physical occupation was a direct confrontation between art and fossil capital. While conservative media and Shell viewed the project as vandalism masquerading as virtue, focusing on the hazardous nature of the act, the artwork's impact was swift and global.[2] The backlash only enhanced its visibility. Online, the image of *Butchered* went viral within hours, circulating across platforms as both meme and martyrdom.

With Kapoor and Banner, art no longer comments on environmental destruction from a safe distance, it infiltrates the machinery of harm. The installations operated within the media economy as a kind of "image terrorism" (in the most tactical sense): a spectacle that compels acknowledgment, that restages the industrial sublime as a crime scene. In the age of social media where a viral post can amplify a message beyond borders, buzzy interventions can propel Greenpeace's mission to new levels.

ART AS A GLOBAL CATALYST

Greenpeace's collaborations with artists have the power to reach wide audiences and share environmentalist messages in a far more personal and impassioned way than science can alone. The artworks the nonprofit has supported call for a reckoning: with complicity, with grief, with latent revolutionary desire. The visuals weaponize affect, making the global intimately proximate. These gestures are performances as much as they are protests. Greenpeace has not merely "used" art, it has shown how art can amplify a message and act as a form of cultural resistance. The museum is too slow, too sealed, too polite. Climate politics requires intervention, occupation, and *détournement*.

Artists who collaborate with Greenpeace step into a zone where aesthetics can meet risk, sometimes including legal and physical. Their work is sited in conflict, becoming like a time-based performance whose duration is often only as long as the target of its message will allow. If the climate crisis is a crisis of imagination and a failure to see alternative futures, then art remains one of the few tools we have capable of prying open the possible.

Annabel Keenan

Annabel Keenan is a New York–based writer covering contemporary art and environmental sustainability for publications including *The New York Times*, *Financial Times*, and *The Art Newspaper*. She is the author of *Climate Action in the Art World: Towards a Greener Future* (Lund Humphries, 2025). She has held roles in museums worldwide, including The Morgan Library and the Victoria and Albert Museum. She holds a BA in art history and Italian language from Emory University and an MA in decorative arts, design history, and material culture from Bard Graduate Center.

Notes

1 Jeffrey Kastner and Brian Wallis, eds., *Land and Environmental Art* (Phaidon, 1998), 16.

2 A Shell spokesperson called the act "extremely dangerous" trespassing. Damien Gayle, "Huge 'Butchered' Artwork Installed on North Sea Gas Rig by Greenpeace Activists," *The Guardian*, August 14, 2025, https://www.theguardian.com/environment/2025/aug/14/greenpeace-activists-install-huge-artwork-by-anish-kapoor-on-north-sea-gas-rig.

Illustrations

A Nicolás García Uriburu, *Coloración*, Canal Grande, Venice, 1968, color photograph, © Fondacion Nicolás García Uriburu

B Nicolás García Uriburu, *Basta de Contaminar*, 1999, color photograph, © Fondacion Nicolás García Uriburu

C Artist Fiona Banner aka The Vanity Press, *Klang Full Stop* outside DEFRA in London, 2020, © Chris J. Ratcliffe / Greenpeace

D Anish Kapoor, *Butchered*, installation photo, 2025, © Greenpeace

WHERE DOES WASTE BEGIN?
WHERE DOES IT END?

A CONVERSATION WITH THE "CRITICAL FRIENDS"

Christina Danick
Peter Emorinken-Donatus
Michael Griff
Roman Köster
Eva Rost

Michael Griff (MG):
We invited you to become part of the Critical Friends, a committee that will be in place for a year, advising the museum as it develops *Waste: An Exhibition About the Global Routes of Rubbish*. You all come from different areas. Why is your work focused on waste?

Peter Emorinken-Donatus (PED):
Coming from the so-called Global South, I know exactly the problems rubbish can cause, especially rubbish you haven't actually produced yourself. In this respect, waste is an issue that I not only tackle politically—by asking, "How can we rectify our consumerist behavior? How can we improve waste disposal?"—but also concern myself with in my day-to-day life. For me, it is an existential issue. I am particularly interested in the displacement of waste, partly because it is very much connected with colonial history.

Roman Köster (RK):
I engage with waste primarily from a historical perspective. When I focused on the topic for the first time over fifteen years ago, I was particularly interested in the side of it relating to economic history—i.e., the question of how waste has developed into a service industry worth billions. Today, I am interested in two issues: What is the actual source of the major environmental problems connected with waste? How did they arise? The other important question is, What do people regard as useless, as dirty, as worthless? How has the definition of rubbish changed over the years? And what does our definition of it reveal about us as a society?

Eva Rost (ER):
As a preschool teacher, I have a particular eye to how we deal with waste at the day-care center. To start with, the focus was on plastic. I only really noticed this when I went to Ghana and saw German packaging waste lying on the beach there. I also realized then that it is not just people who are afflicted by waste but animals too, which is always a good place to start when you're working with children in an educational setting. Like Peter, I've since been paying attention to the footprint I leave behind, and how I can do things differently in my life.

Christina Danick (CD):
In our meetings, we talked a lot about what waste actually is. How would you both define it, and what are the difficulties involved in arriving at a definition?

PED:
Naturally, this is an important question: Where does waste begin? Where does it end? What is junk for me—for us here in the Global North—is worth a mint in the Global South, helping people survive. Waste is a product of uncontrolled consumerism, but it is always being

created, even in nature. It becomes a real problem when it spreads uncontrollably. If it devastates our biodiversity, damages people's health, or destroys livelihoods, then I would call it waste.

RK:
I don't think waste is something you can really define. You might say that it is what people deem to be rubbish. But that is very relative. It is viewed differently in different parts of the planet, depending on how rich or poor people are, what ideas they have about dirt and cleanliness, and so on. Based on this logic, there are no objective properties that make something trash, but at the same time we all know what it is. We throw things away without really thinking about it. And I find this dynamic interesting. The obvious fact that our everyday lives generate rubbish, the way we live day to day—this isn't something we give much thought to. On the other hand, trash creates major environmental problems in a large society.

ER:
The issue of coming up with a clear definition is also mirrored in my day-to-day work. For example, we had a much bigger problem than recycling at the day-care center—and that was food wastage. When I started getting involved in the issue of waste, I often wondered why it is we chuck away so much food. This is actually a key question for people who want to act in a sustainable way. Day-care centers aren't allowed to reuse leftovers after a certain period of time—this is down to laws designed to protect against disease. And at the same time, it is perfectly alright to throw away ten kilos of food every day.

MG:
Your answers are indicative of how power relations are reflected in our dealings with waste. Roman, how do you convey to students that waste is something worth focusing on, even if it can't be clearly defined?

RK:
The goal should be to understand why any clear definition of waste is an impossibility. This is because there are so many factors governing it. It is a question of poverty and wealth, geographical differences, power gradients, and a great deal more. It is not about saying, "This is waste," but instead asking, "Why do we struggle so much with it? Why is it so difficult to define?" Students should understand at a fundamental level that the world isn't easy to explain, that things are not as clear-cut as they might seem at first glance.

PED:
A few days ago, I had an intense discussion about rubbish in parts of the city with a low-income population. Most of the people involved in the discussion group were middle-class and wealthy and were complaining, for example, about the trash on the streets of the Chorweiler borough of Cologne. I asked them about the

waste they produce themselves on the basis of their wealth. I think it was only then that they became aware of the scale of it. It is important to check yourself—not only for individuals but for the Global North, which could immediately opt out of fossil fuels. But that is not happening. Instead, the burden is being shifted to the countries of the Global South, which in turn produce oil so that they can pay off their loans in the Global North. This kind of politics is two-faced and does nothing to help tackle the environmental issues together.

CD:
This is a key point, because you have to be able to afford to dispose of waste and push it out of your awareness. For those who cannot afford waste management or live in areas where the structural conditions do not allow it, rubbish will continue to be an issue.

RK:
I would like to note, though, that the perception of a place as full of rubbish or particularly dirty is connected with a general sense of crisis in society. I have had many conversations in the last months where people have told me, for example, how dirty Berlin and other cities are, which doesn't tally with my personal perception at all. I wonder whether this sense of there being rubbish everywhere isn't rather a symptom of an overall perception of crisis.

MG:
Are there historical instances of the perception of waste being instrumentalized like this?

RK:
Crudely put, one might say that it is part of human nature. That people see themselves as clean and others as unclean—in other words, they perceive others as worse than themselves: That has always been the case—there is evidence of it in ancient history. Something has changed, though, in the modern era: People have started to use dirt for their own ends and make a strong connection between it and morality. Which does not really make sense per se. Why should someone who does not wash their feet, say, be a bad person? Or someone who does wash them, a good person? It is only in the modern era that cleanliness has become associated with order, with people's moral qualities, with certain ethnicities, with social groupings. In this respect, there is nothing surprising about what is happening right now, the fact that there are such hard-fought debates about urban cleanliness at the local level—there are many historical examples of this. I think it is significant that it is currently intensifying and running in parallel with all the other crises confronting us. My contention would be that urban cleanliness hasn't really got any worse in the last ten or twenty years.

PED:
At the same time, we should not forget the role that colonial history

plays here. Let us take the so-called re-education of Indigenous peoples in colonized countries, whose traditional practices were banned, including waste recycling, and forcibly replaced by those of the colonial powers, as exemplified by their mode of waste production. Their own practices were typically in harmony with nature. The *white* man compelled these people to abandon their sustainable method of waste recycling in the name of a purported "civilization." So we shouldn't be surprised that we now find ourselves in this profound crisis affecting our environment and climate. This has to do with the so-called educational measures, which we term cultural genocide in the language of pan-Africanism. This means that overcoming the problem in the future will require an analysis of history.

ER:
That's right—because racism legitimizes the idea that waste can be shoved somewhere else.

CD:
We tried to take a closer look at the routes taken by waste as part of our research work for the publication. For example, European countries sell vast quantities of plastic waste to Southeast Asia, and some 15 to 30 percent of all Europe's waste trade operates illegally. What is your take on the relationship between political regulation and individual responsibility?

RK:
For over thirty years, people have been trying to find political ways of regulating the waste trade and illegal practices in dealing with trash, yet the situation is still not clear: Are people looking away? Do there need to be more controls in place? Can it actually be controlled at all? Or does the fact that we find it so difficult to get on top of illegal waste practices actually come in handy, because it makes it easier for society to displace the toxic substances, residues, and problematic waste we produce? This is sometimes difficult to determine, but it is important to ask the question.

PED:
I think environmental protection is a privilege. The system of separating waste we have here in Germany, in Europe, is one aspect. Germans are proud of it at this point—it is part of the culture. An emancipatory measure has turned into a legal obligation. But what about the majority of the world? In the places where the bulk of the global population live—these people don't have the resources to actually establish a system like this. Compared to people who live elsewhere, I'm privileged in this regard. We have to constantly do this check, because without going through this process of self-reflection, we can't really talk about waste at all.

ER:
Not only do companies need to be held accountable, the state must too.

Do you know anything about what Germany is doing politically to make this happen, what decisions are being made on that front?

RK:
Yes, there are a number of international accords. The 1989 Basel Convention has, I believe, been expanded at different times. There have been various summit scandals that have grabbed the headlines. And these are actually very damaging for companies if they are indeed caught out. The sanctions imposed on the company Trafigura are a good example of a really tough punishment being meted out to a company for illegal practices. The export of plastics to Asia is a business model. There are certain recycling quotas here, which can be fulfilled more easily if you have the stuff officially recycled in Asia. Certain Asian countries—prior to 2018, this was primarily China—take in the plastic waste and then make money from it. This is certainly not the ideal solution, but it is, in principle, a consequence of our recycling policy.

CD:
An EU regulation on shipments of waste was passed in 2024—the rules will apply in 2026. It includes a ban on the export of plastic waste to countries that aren't members of the OECD. On the basis of this, plastic waste may no longer be shipped to certain countries in Southeast Asia. But what will probably happen then is that the waste will be shipped to Turkey, which is already the case now to some extent. Today, Turkey is a massive transshipment point for waste and scrap, much of which is sold on and shipped from there.

Eva, is waste already a focus in environmental education—for younger age groups, in particular? If so, how is knowledge taught here, and what takeaway lessons can be given to young people?

ER:
I think other forms of discrimination also play a role here, including adultism. This is a form of discrimination against children. It derives, among other things, from a desire to protect children. In my view, this is also why there is no discussion with children about colonialism.
In my work, I have always stood by the idea that children are just junior people. Most importantly, I'm repeatedly struck by how you can discuss all kinds of things with children. They have an intrinsic need for other people to be okay. With children, there is never any problem talking about waste or recycling, and why people should do it.
The fear and resistance have always come from adults.

I think that we, as a society, need to get back to the idea that it is fine to make mistakes and not know certain things. But how are we to act in concrete terms when we know the consequences of the waste trade and overconsumption? Do we keep ignoring the issues because they require us to change things? Or do we open up to them because we all want the Earth to be here a while longer?

PED:
I don't think education is a one-way street from parents to children anymore—often it goes in the other direction. That is what I've observed when it comes to the environment and sustainability. We can build on this capital. These issues resonate with children and teenagers, and they quickly respond to them. We can see this in relation to Fridays for Future, a movement that is no longer just limited to the youth generation now. When young people's commitment gets an entire society moving, real change occurs.

Peter Emorinken-Donatus, born in Nigeria, is a freelance journalist and education consultant. He is known as a long-standing pan-African environmental activist and declared opponent of the Shell Corporation, and is considered one of the most prominent voices from the Global South for environmental and climate justice in Germany, where he has lived for over thirty years. He is a member of the Greenpeace supervisory board and the initiator, co-founder, and spokesperson of Bündnis Ökozidgesetz, founded with the aim of criminalising ecocide. Together with a number of experts from the Global South living in Germany, he founded a BIPoC think tank: Care & Repair – Decolonial Think Tank for Environmental Justice.

Roman Köster is an economic and environmental historian, whose research is focused on the history of economic crises and of waste. He is an associate professor (PD) at the University of the Bundeswehr Munich. His book *Müll: Eine schmutzige Geschichte der Menschheit* (C. H. Beck, 2023) was nominated for the 2024 German Non-Fiction Prize.

Eva Rost (gruenekitas.com) is qualified as a state-certified preschool teacher and trained with Diakonie (STUBE Nord) Hamburg as a multiplier disseminating solutions for tackling polycrises. Prompted by Germany's shift to the right, she has stopped working in kindergartens for the time being. She sees herself as an educator in decolonial art with a varied focus. Her previous projects include a collaboration with the European Youth Education Center (EJBW) in Weimar, involvement in the anti-racism project "Film Macht Mut," and work as an assistant theater pedagogue and consultant for "Digger, denk mal nach ..."—a theater project with high school graduates organized by the Oldenburg chapter of the German Red Cross (DRK), Kreisverband Oldenburg-Stadt e.V. She is involved in a small political party with an anti-racist agenda and stood for election at state level as a Hamburg representative. She appears under her own name on LinkedIn and is part of the pool of speakers for the Initiative of Black People in Germany (ISD).

Ana Alenso, Obsolete Swing, 2026, © Ana Alenso

AKWASI BEDIAKO AFRANE

TC-2000, 2026
Mixed media installation, incl. cables, PCBs, computer hardware, ca. 250 × 600 cm (height × diameter); New production as part of the exhibition

DIETRICH ALBRECHT

Es wird sehr viel vom Umweltschutz geredet!, n. d.
Offset on paper, 10.2 × 14.2 cm; Collection Museum Ostwall at the Dortmunder U

ANA ALENSO

Obsolete Swing, 2026
Part of the series Mad Rush, 2022–ongoing, E-waste, scrap metal, used golf clubs, cable ducts, metal pipes, belt conveyors, microscope, monitors, color photographs, hoses, minerals, water, earth, variable dimensions; On loan from the artist, new production as part of the exhibition

ARMAN

Untitled, 1972
Rubbish cast in polyester, 100 × 50 × 10 cm; The Arman Marital Trust, Corice Arman

KARIMAH ASHADU

Brown Goods, 2020
Single-channel video installation, 12 min.; On loan from the artist and Sadie Coles HQ

KADER ATTIA

Los de Arriba y Los de Abajo, 2015
Metal structure, roller blinds, metal fencing, trash, ca. 200 × 1500 × 290 cm; On loan from the artist and Galerie Nagel Draxler Berlin/Cologne/Meseberg

Here and following pages: Allan Kaprow, Andrew Glantz, Transfer (for Christo), 1971 (see exhibited works)

CÉSAR BALDACCINI
(aka "César")

Untitled, 1959
Painted automotive metal parts, welded together with industrially manufactured steel sheet metal boxes, 43.5 × 33 × 17 cm; Collection Museum Ostwall at the Dortmunder U

HICHAM BERRADA

Carte mère #11, 2020–23
Copper, aluminum, silver, lead, acrylic glass, illuminated, 37 × 28 × 5 cm; On loan from the artist and Mennour, Paris

Carte mère #14, 2020–23
Copper, aluminum, silver, lead, acrylic glass, illuminated, 37 × 28 × 5 cm; On loan from the artist and Mennour, Paris

Carte mère #19, 2020–23
Copper, aluminum, silver, lead, acrylic glass, illuminated, 37 × 28 × 5 cm; On loan from the artist and Mennour, Paris

Carte mère #20, 2020–23
Copper, aluminum, silver, lead, acrylic glass, illuminated, 37 × 28 × 5 cm; On loan from the artist and Mennour, Paris

Carte mère #21, 2020–23
Copper, aluminum, silver, lead, acrylic glass, illuminated, 37 × 28 × 5 cm; On loan from the artist and Mennour, Paris

H. R. DECKER

Nur dumme Fische sterben in verschmutzten Flüssen, 1974
Artist's book, offset and silkscreen on paper, 29.9 × 20.9 cm; Collection Museum Ostwall at the Dortmunder U

Umdenken ist Umweltschutz, ca. 1972
Offset on paper, 29.5 × 20.9 cm; Collection Museum Ostwall at the Dortmunder U

MARK DION

Spider Monkey, 2013
Monkey skeleton, tar, glass, metal, porcelain, plastic, paper, ceramics, string, fabric, cork, copper, in museum display case on wooden shipping crate, 163 × 132.5 × 102 cm; Collection Museum Ostwall at the Dortmunder U

ELLIE GA

Gyres, 2019
Single-channel video installation, 39 min.; On loan from the artist and Bureau, New York

JAN HENDERIKSE

Untitled, ca. 1960
Plastic box with screws, plastic figures, axles from toy cars, bottle corks, thumbtacks, shards, light bulb, plastic springs, 108 × 178 × 4.1 cm; Collection Museum Ostwall at the Dortmunder U

NANCY HOLT

Sky Mound – Sun Viewing Area with Pond and Star Viewing Mounds, 1985
Graphite on paper, 597 × 120.7 cm; Holt/Smithson Foundation

Sky Mound – Sun Viewing Area as Seen from the New Jersey Turnpike, 1985
Graphite on paper, 59.7 × 120.7 cm; Holt/Smithson Foundation

Sky Mound – Moon Viewing Area, 1985
Graphite on paper, 53.3 × 113 cm; Holt/Smithson Foundation

Sky Mound – Sunrise on the Spring and Summer Equinoxes, 1985
Graphite on paper, 59.7 × 76.2 cm; Holt/Smithson Foundation

EBRUARY

Original site plan for Sky Mound, 1985
Blueprint, 88.9 × 104.1 cm; Holt/Smithson Foundation

Bird's-eye view of the original site for Sky Mound, 1984, Color photo, Digital copy as wall print; © Holt/Smithson Foundation, licensed by Artists Rights Society, New York

Sky Mound under construction, 1991, Color photos, Digital copy as wall print; © Holt/Smithson Foundation, licensed by Artists Rights Society, New York

ALLAN KAPROW

Transfer (for Christo), 1968
Offset on paper, 27.9 × 55.4 cm; Collection Museum Ostwall at the Dortmunder U

Allan Kaprow (artist), Andrew Glantz (photographer): Transfer (for Christo), 1971, photograph, silver gelatin DOP on Baryta paper, 107 × 112 cm, digital copy as wall print; © Allan Kaprow Estate, Courtesy of Hauser & Wirth and Andrew Glantz, Zenith Design, Digital copy: © Staatsgalerie Stuttgart

KRIŠTOF KINTERA

Postnaturalia, 2016/17
Cables, plastic, PCBs, computer hardware, dimensions variable, ca. 1200 × 800 cm; Kunsthalle Praha

FRANCOIS KNOETZE

Core Dump – Kinshasa, 2018
HD digital video, 12:42 min.; Collection Museum Ostwall at the Dortmunder U

Core Dump – Dakar, 2018/19
HD digital video, 12:17 min.; Collection Museum Ostwall at the Dortmunder U

Core Dump – Shenzhen, 2019
HD digital video, 12:17 min.; Collection Museum Ostwall at the Dortmunder U

Core Dump – New York, 2019
HD digital video, 12:47 min.; Collection Museum Ostwall at the Dortmunder U

CHRIS REINECKE

Konserve, 1968
Tin can, oil paint, height 12 cm; Collection Museum Ostwall at the Dortmunder U

Golden Delicious, 1968
Brown paper, corrugated paper, blue paint, stickers, glued into plastic film, 56 × 88 cm; Collection Museum Ostwall at the Dortmunder U

DIETER ROTH

Die Welt mit Kram drauf der verdampft, 1966
Etching on paper, 75 × 56 cm; Collection Museum Ostwall at the Dortmunder U

HA SCHULT

Situation Schackstraße, 1969/70
Four-part series, Silkscreen print on paper, 48 × 68 cm; Collection Museum Ostwall at the Dortmunder U

Biokinetische Olympia-Situation, 1972
Wooden table, sand, earth, metal, plastic, paint mold, fungus and bacterial cultures, 88 × 174 × 206 cm; Collection Museum Ostwall at the Dortmunder U

TEJAL SHAH

Between the Waves, 2012
Five-channel video installation, Various lengths and loops; On loan from the artist and Project 88